Rend the Heavens …
and Come Down

Bill McGreehan

KJV and Use of BCE and CE

a. Unless otherwise designated, Scripture quotes are taken from the King James Version (KJV) and not the original 1611 King James English which sounds excessively archaic.

b. Word meanings provided are based on original Hebrew and Greek texts.

c. Supposedly for religious neutrality, since the 20th century secular publications have been increasingly adopting the 17th century scientific convention BCE (Before Common Era) and CE (Common Era) as alternatives to what had been common with BC (Before Christ) and AD (Anno Domini, or Year of Our Lord)—In my opinion, both these conventions are arbitrary and designed to satisfy no one, so for the sake of bridging that divide, this book will abbreviate BCE to represent "Before the Christian Era" and CE for "Christian Era."

Oh that thou wouldest rend the heavens,
that thou wouldest come down.

Isaiah 64:1a

Dedication

In the days leading up to publication of *Rend the Heavens,* a life was cut short but not without first casting a long shadow. Charlie Kirk (1993-2025), founder of Turning Point USA, embodied the themes of this book to a greater degree than anyone in recent memory. His disproportionate influence on the younger generation and the culture have helped set the stage for God's closing scene of history described here. This book is dedicated to the legacy of Charlie Kirk and that final drama of Christ's return.

Through dialogue, Charlie Kirk exploited the resistance of opponents to press the truth, always striving to make the problem part of the solution. Rather than accumulate followers, he made disciples, who make disciples, who make disciples…. Charlie viewed his life, first and foremost, as an investment in the furtherance of the gospel of Jesus Christ. And Jesus assures, all your investments are safe and earning dividends.

Yet no human exploit could tip the scales of history were it not for God's window of revelation making Himself known in the heart and mind of the believer. That confidence is the promised rest hinging on the completed work of the cross by Jesus Christ Himself.

Charlie Kirk dedicated himself to that principle, knowing this generation faces an intolerable alternative; receiving the ultimate consequence of, "What you demanded."

Acknowledgements

Many of the stories and teaching material for *Rend the Heavens* were developed over several decades, but the final concept for the book would not have come together during the past year without the help of several key contributors.

First of all, my wife Debby has been patient with the creation of this book, as with all of my projects the past fifty-three years. Regularly Debby petitions, "Can I be your next project?"

Appreciation goes to editors Scott Bueling and Michelle Cox for their effort and insistence that the wording be simplified and readable. And my indebtedness to Chris Taylor (www.seraphimchris.com) for the outstanding book cover design. Also, Richard Eijkenbroek (www.Xquissive.ai) made a major contribution to *Rend the Heavens*. Richard's artistic work created most of the illustrations and signposts based on my rough concepts.

Of course, none of this book would have materialized were it not for the work of God these past seventy-seven years, in particular the miraculous way Jesus Christ and the Holy Spirit have change my life, family, and the lives of others. Fifty-five years ago, while at Penn State University, I declared, "No matter what you believe, God accepts you." My roommate, Jay McCarroll responded, "That's not true, Jesus Christ is the only way to know God." To that I retorted, "I don't believe that." But whatever Jay had, I knew I needed. To God be the glory.

Contents

Introduction

BEFORE OUR COLLEGE spring break back in the 70s, our team from Penn State Intervarsity Christian Fellowship (IVCF)—now International Fellowship of Christian Students (IFCS)—had prepared for a bus trip to evangelize Fort Lauderdale Beach.

Just a few months prior in November, 1970, I committed my life to Christ through the witness of my roommate, Jay McCarroll, who was leading the IVCF group on campus. I prayed and signed a commitment on the back of a Jack Chick cartoon witnessing tract. That night, the following assurances came to me:

- My sins are forgiven.
- Jesus is in my life.
- He is the Son of God and the only way to know God.

Those realities arrived as through a window in heaven, louder than an audible voice. Also came inner confidence and understanding that Jesus would soon return. Though having previously read through the Bible, it now came alive with truth and meaning, combining into a clear message.

As a regular weekly attender of Catholic Mass at Penn State, eventually the resident priest requested I write a weekly, one-paragraph introductory thought to preface his sermon. Newly enlightened by a personal relationship with God and Scripture, my opening message became clearly evangelistic.

After a few weeks of reading my submissions, he chastised, "You can't say those things. I need to meet with you." At that meeting, and upon receiving my written list of doctrinal concerns, he finalized his opinion of my strongly-held beliefs, "I don't think there is any hope for you."

By spring break that March, I discontinued my master's degree studies in Aerospace Engineering and decided to join the IVCF evangelism group in Fort Lauderdale. But rather than travel with the rest spending money on bus fare to Florida, I elected to hitchhike and witness along the way.

Fifty years ago, this didn't present a significant risk and had become a popular mode of getting around. As I set out on my hitchhiking journey to Fort Lauderdale, the first day was quite uneventful, with numerous offers of rides. Late in the afternoon on this particular day, a Baptist minister had offered a ride—and some encouragement I had not anticipated needing. But God already knew about the encounter I would experience the next day...

Hitchhiking to Fort Lauderdale, Florida

Chapter One - On The Road

IN MARCH, 1971, at 6:00 am, a hazy sun had already peered over the backroads on the North Carolina, South Carolina border. After a restless night in my sleeping bag, I woke late, quickly reassembled the backpack, grabbed my cowboy hat, and picked up my sign before quickly strolling back to the roadside.

The previous days' hitchhiking south from Penn State University had covered close to six hundred miles. Promptly this morning, the first driver to come along pulled over and offered a lift.

Not in the best condition, his vintage Thunderbird (probably 1957) exuded a sun-bleached patina and junkyard scent, but it looked road-worthy. Throwing my backpack on the tarp behind the bucket seats, we engaged in conversation.

The driver appeared as rough as his T-bird. His gaunt face and the deep scar traversing his right arm from wrist to elbow fit his abrasive demeanor. Though the conversation at first remained casual and absent any red flags, as the driver recounted a recent release from prison for manslaughter and a "system" that did him wrong, he grew increasingly agitated.

Methodically he recounted his collection of prison memories and claimed that two pitchers of beer the night before had left him with a throbbing hangover.

Ride with an Ex-convict

Eventually finished with his rant, he burst into a rage and blurted a question, "Do you know what I do when I see anything living along the road?" Without waiting for a response, he glared over, reached back between the bucket seats and pulled away the tarp to reveal a shotgun and rifle. Answering his own question he added, "I *kill* it!"

With clenched jaw and gripping the wheel, he stared straight ahead and almost immediately found a dirt road leading off the main highway. Exiting, he took us speeding into the woods, all the while refusing to react to my occasional glance his way.

With a premonition of his intent, my heart felt as if it was pounding loudly enough for him to hear. With no other recourse, I decided the only tactic was to share my story of what God had done in my life. Experience had taught, when faced with danger, always make the problem part of the solution.

As the hardened convict sped along the dirt road, I began explaining how God had changed my life. I told him how I'd decided to take this trip, with a plan to witness for Christ on the beach in Fort Lauderdale. Not long after I'd begun talking, the driver mellowed, slipped back from Mr. Hyde into Dr. Jekyll, and resumed talking.

Slowing the car, he turned around, drove back, and the T-Bird exited the woods onto the highway. He drove us south to Savannah, Georgia, where he wished me well and departed. Not until he drove off did I realize how the Spirit of God and the spoken Word had intervened, overruling his intentions, and providing protection for me.

Arriving the next day in Fort Lauderdale and connecting with the group from IVCF, someone introduced me to a team of youth from World Ministry Outreach (WMO), who were also evangelizing on the beach. After living at WMO facilities for a few days, I sensed a calling to work with them, in Fort Lauderdale, and in the future.

Eventually WMO included seven outreach centers, one of which was a mission center in Port-au-Prince, Haiti. In August, 1971, I traveled to Port-au-Prince with WMO for a crusade, and stayed and worked there for several years. My wife Debby and I met working with WMO in Haiti, where we ministered in full-time missions and later directed the Gainesville, Florida center until late 1975. The events in Haiti will be a side trip I'll take you on down the road.

During our over four years with WMO (at the time of the so-called Jesus Revolution) we saw hundreds come to Christ—mostly youth—and innumerable miracles that transformed lives and the spiritual landscape. Documenting those events and the impact on the kingdom of God would take a bookshelf.

By 1975 WMO began to experience many of the issues characteristic of ministries established during that period. In God's purpose, the seeds of young believers scattered across the nation and around the world. For Debby and me it had been a memorable road trip and opportunity for influence.

World Ministry Outreach, Fort Lauderdale, Florida

Looking back to the winter of 1971 while still enrolled at Penn State, a campus ministry conducted an outreach presentation on scientific explanations of Bible miracles. Ministry promoters for the event created sufficient interest to pack the lecture hall. The guest speaker launched into an explanation of familiar Bible stories characterized by miraculous events. Describing each, he offered a plausible, natural explanation to make the story more believable.

One memorable example included a now-common, alternative description of the Red Sea crossing from Exodus 14:21-31. In this case, the parting of the waters and passage of the children of Israel was suggested to have been facilitated by a wind storm and tidal movement in a shallow basin of the Red Sea.

No attempt was made to explain how Pharoah's army drowned making the same crossing. In another instance the speaker suggested Jesus may have walked on the Sea of Galilee by taking advantage of strategically placed steppingstones on a sand bar.

Though the presentation ended with an invitation to faith, the approach could hardly be characterized as faith-building—quite the contrary. Efforts to naturalize biblical accounts to make them plausible, or attempts by skeptics to dismiss the supernatural, both undermine scriptural authority and scientific rigor. By His Spirit, God alone gives the inner person assurance to the truth of His Word.

In 1971, Don McLean released his song "American Pie," a profound musical commentary on the dark side of culture and rock-and-roll. The song's lyrics take an eight-and-a-half-minute journey through scenes along the American cultural roadway of the 1960s, with the chorus always returning back to the levy where the riverbed has dried up.

McLean's wandering imagery offers a blueprint for chapters ahead. My book will look back on events through a similarly troubled landscape, though focusing on a hopeful destination. Rather than the desolation of McLean's parched and lifeless song-ending verses, we'll talk about the promised appearing of our resurrected Jesus Christ.

For decades McLean refused to offer meaning to the lyrics, only responding cryptically that the song's popularity meant he would never have to work again. Finally in 2015 at age 69 (when the original written copy of "American Pie" sold at auction), McLean observed that the song was about society heading in the wrong direction. He mourned a lack of poetry and romance, finishing his portrayal with a last verse that laments the exit of the Father, Son, and Holy Ghost—meaning the music had died.

Of the many verses he had written, McLean chose not to record a final one that could have left the listener with a much different feeling. His originally planned ending was scrapped because he thought its hope did not reflect the mood that the rest of the song had created. Those unrecorded words spoke of him standing alone and then kneeling to pray. After he promises to relinquish all if only the music would live again, the lyrics report a great hope he receives from divine assurance that the music would be reborn.

Those left-out words accurately reflect what God does from His vantage-point. Don drew close to getting the script right, but hesitated short of hope. Each verse of "American Pie" ends with a springboard reference to the 1959 plane crash that took four lives, including McLean's music idol Buddy Holly, along with J.P. Richardson, Ritchie Valens, and the plane's pilot Roger Peterson. Unlike the bleak ending the song settles on, God's plan always fosters rebirth, as related in these stories. The river was not dry, just following a new course.

Similar to Don McLean's "American Pie," popular culture commonly presents life's flipside to the eternal story, like meaningless threads on the backside of a tapestry. Our society in the 2020s mirrors the same emptiness of the 1960s and early 1970s that struggled to identify winners and losers. At that time, God by his Spirit birthed the Jesus Revolution, setting the stage for something yet to come.

Over fifty years have rushed by since then, and during each, commentaries have tried to sort out the cultural landscape. Regrettably, most view the scene cynically from the backside, or with empty optimism, rather than as transforming events according to God's purpose.

In Genesis 22, God had Abraham travel three days with his son, and stand on Mount Moriah with instructions to sacrifice his promised son Isaac. Carrying the sacrificial wood for the fire, Isaac questioned his father about where the sacrifice would come from. Abraham assured his son, "God will provide himself a lamb."

On the mount, the angel restrained Abraham from acting to sacrifice his son. Then Abraham spied a ram with its horns caught in the thicket and pronounced the place Jehovah-Jireh, meaning "the Lord provides."

Abraham and Isaac on Mount Moriah

Providence (to provide) implies seeing the future and planning accordingly. From Mount Moriah, Abraham looked over the hill that was to become a place of sacrifice where King Solomon would build a temple (2 Chronicles 3:1). With Jesus as head cornerstone (Psalm 118:22), that temple overlooked Calvary's fulfillment when God provided Himself a Lamb in Jesus.

Like the good Samaritan (Luke 10:25-37), Jesus did not condemn and pass by on the other side of the road from a fallen world. Rather, like Isaac, he shouldered the wood, faced the cross, and provided a solution to humanity's sin problem. John revealed more about this in his divine vision of end times to come:

> And one of the elders saith unto me, Weep not: behold, the Lion of the tribe of Juda, the Root of David, hath prevailed to open the book, and to loose the seven seals thereof. And I beheld, and, lo, in the midst of the throne and of the four beasts, and in the midst of the elders, stood a Lamb as it had been slain, having seven horns and seven eyes, which are the seven Spirits of God sent forth into all the earth. And he came

and took the book out of the right hand of him that sat upon the throne (Revelation 5:5-7).

God continues to prepare the world for a future decisive shift—by establishing and removing cultures, to set up the final act of recorded history that will come with Jesus's sudden reappearing:

And I beheld when He had opened the sixth seal, and, lo, there was a great earthquake; and the sun became black as sackcloth of hair, and the moon became as blood; and the stars of heaven fell unto the earth, even as a fig tree casteth her untimely figs, when she is shaken of a mighty wind. And the heaven departed as a scroll when it is rolled together; and every mountain and island were moved out of their places. And the kings of the earth, and the great men, and the rich men, and the chief captains, and the mighty men, and every bondman, and every free man, hid themselves in the dens and in the rocks of the mountains; And said to the mountains and rocks, Fall on us, and hide us from the face of him that sitteth on the throne, and from the wrath of the Lamb: For the great day of His wrath is come; and who shall be able to stand? (Revelation 6:12-17)

In a parallel description as part of his Mount of Olives passage, and in response to the twelve disciples requesting details of his return, Jesus declared:

Immediately after the tribulation of those days shall the sun be darkened, and the moon shall not give her light, and the stars shall fall from heaven, and the powers of the heavens shall be shaken: And then shall appear the sign of the Son of man in heaven: and then shall all the tribes of the earth mourn, and they shall see the Son of man coming in the clouds of heaven with power and great glory. And he shall send his angel with a great sound of a trumpet, and they shall gather together his elect from the four winds, from one end of heaven to the other (Matthew 24:29-31).

These two accounts of dramatic earthly and heavenly events have no equal in human history. As such, trying to fit them into traditional patterns of thinking makes no sense. Both stories reflect changes to our sun, moon, and stars, as well as unparalleled disruption on earth. Jesus added the following assurance that the events he previewed would happen:

> Verily I say unto you, This generation shall not pass, till all these things be fulfilled. Heaven and earth shall pass away, but my words shall not pass away (Matthew 24:34,35).

The expression for the day of the Lord in Revelation 6 employs the phrase "great day of His wrath." By means of the day of the Lord (in Hebrew, Yowm Yahweh, or Yowm YHWH), God represents a key theme of the Old and New Testaments as it relates to the promised return of Jesus Christ—though the message rarely receives adequate attention. Past history, current events, and future promises all weave together into a three-fold cord at the completion of the age.

In Revelation 5, a roll (scroll, or book) appears with seven seals. Upon opening the Sixth Seal, Christ the Lamb illustrates God's purpose in the created order of the heavens and unfolds his return to catch away believers before impending judgment.

The Revelation of Jesus Christ by the Apostle John brings to completion numerous stories of God's plan for creation, as well as providing a roadmap of future hope.

Much of the account of creation, the fall of man, and promised redemption leads to the final conflict between the King and adversary. The descriptions in future chapters of scriptural themes, do not seek to reveal anything new—only to add clarity to what is recorded. Scripture and history demonstrate the outcome.

1960s Sidman Home

Chapter Two - Sidman

MY SOPHOMORE YEAR at Triangle Area High School (TAHS) began with a commitment to improve my below-average grades and hopefully even excel academically. But rather than succeeding as planned, the next year became transformational in ways I hadn't expected.

While sophomore years have the reputation as a time of outrageous and painful lessons, mine involved making a powerful telescope and becoming fascinated with rocketry, science, and space. From the consumption of multiple science magazine subscriptions (courtesy of a generous aunt), faith in the natural order, logic, and engineering left its impression—and a sense of foreboding.

As is often the case with the science culture, much of my reading material scoffed at the supernatural. I soon posed a silent challenge for God to make Himself known to me by knocking off a certain tree branch. Seeing no response, it seemed natural to embrace atheism. Though I was regularly serving as a Catholic altar boy, for me the idea of God and Jesus lost relevance.

Upon informing classmates about my rejection of faith, two young ladies conspired to become my informal intervention committee. They spent weeks hammering me on matters of faith, occasionally presenting arguments coached by their parents.

It took a few weeks but when I finally relented, they rejoiced at my telling them that they had changed my thinking. I had taken a step in the right direction, but only superficially. (By the way, eventually that tree branch did fall.)

Upon purchasing and carrying around H.H. Kohle's *Handbook of Aeronautical and Astronautical Engineering*, an expensive publication by McGraw Hill—and announcing my intentions to pursue a career in aerospace engineering—laughter and comments came from a few classmates, "Your grades would never be high enough to become an engineer."

This was a fair challenge since my few B's and mostly C's were decidedly unimpressive, especially in comparison to the mostly A's achieved by my brother John.

Despite the doubts, income from a newspaper route fed my thirst for formulating and testing rocket fuels. After making a first batch of paper-tube rocket canisters in October 1963, discretion went out the window one day before school when I stuffed a small loaded rocket motor into a coat pocket.

When Mr. Keller left the room for a smoke in the lounge, my informal display to the class resulted in their encouraging, "Shoot it out the window." After a few double-dog dares and even a triple-dog, we opened a window over a field, scavenged a paper-towel tube, and launched the rocket out of sight.

Room 102 Rocket Launch

An unforeseen exhaust blast blackened the window and the classroom filled with smoke. Everyone sprang into action with the ladies spraying their perfumes, the guys fanning smoke outside, and others scrubbing the exhaust smudge off the window.

One classmate stood watch for our teacher and soon the cry went out, "Hurry, Mr. Keller is coming!" A moment later he strolled into the room, seats now occupied and everyone sheepishly waiting for some query such as, "What's burning?" But not a sound. Minutes passed. Nothing. Was he just dismissive—or accustomed to smokey rooms?

A few years later, an accomplice scrawled a senior yearbook entry ending with, "…remember the fun we had in Keller's room, especially the day we launched the rocket." Another from the smoke fan club wrote, "Remember Room 102's rocket launching." A girl signed off with, "Luck in all you do. Especially your bombs and rockets." While touring the 55-yr-old high school at our 50th reunion, we wondered how Mr. Keller did not know. But one of the ladies assured, "He knew, he just didn't want to deal with it."

The diversion of extracurricular rocket projects began taking a toll on my studies as well as caught the attention from neighbors. On my newspaper route a local constable inquired from his porch about who might be setting off explosives in town. Without him volunteering any information, he cautioned, "If I catch you, there will be trouble."

Though I'd committed to becoming an aerospace engineer, by the end of my sophomore year, every subject suffered and geography risked crashing and burning. My final exam remained the only chance to pull out a passing grade. Shuffled across the rows came exam papers that had been scored just days before the close of school. Worse than expected!

Staring at another dismal failed exam, only a "Hail Mary" prayer would do—and it had to be an effective one. Memorably it came out, "God, don't let me fail geography, even if I have to be a priest." At that point a C or D grade seemed too miraculous a request, but upon slowly opening my report card on the bus, there it was: "Geography Final: B."

Was it a mistake? The grading curve? Or … a miracle? At that point it didn't matter. Without a goal for improvement and a plan to get there, a single miracle seemed useless. With only a couple days left the sophomore year, a frenzy of desperation stirred a new goal: straight A's my junior and senior years.

The sudden plan for my last sophomore day became confiscating every junior-year textbook and spending the whole summer locked in my bedroom, reading and working every problem in each book. For good measure, even teachers' additions that included answers were fair game.

Improbable as it was, I felt no alternative remained and the sense grew daily that this would not be some overnight inspiration that would fade. The goal became tangible and the plan more than believable. Soon the realization developed that straight A's were not just inevitable but came with a feeling of divine assurance.

Summer 1964 passed as quickly as all those chapters and books. I was up before sunrise, delivered newspapers, studied an hour, exercised, studied another hour, exercised again, took a snack break, and then went back to the routine until delivering the evening newspapers. At the end of each day, I hid textbooks under the mattress.

By the end of summer break, every problem in each textbook had been completed, checked, and corrected where needed. Most importantly, that routine developed an understanding of every concept, making it second nature to create and work additional problems.

Junior-year classes posed a challenge of expectations since our school's classes were divided into two academic levels: upper and lower. I had always come in at the bottom of the lower group before.

Clawing out scholastically should have been difficult after years of neglected study habits. But that first six-week period resulted in almost perfect scores on all exams. Even extra-point questions seemed a breeze. Going home on the bus with a report card of straight A's, my brother John demanded, "Let me see your report card."

To my emphatic *no,* he chirped, "I bet you have failing grades."

Then seizing the opportunity to drive the stake deeper, John rushed off the bus and into the kitchen where Dad and Mom were waiting for the report-card tribunal. John triumphantly blurted out, "Bill has F's."

Disheartened, my parents let out collective sighs and Dad moaned, "Let me see."

Our family's other three older children usually had great grades. Mine had brought routine lectures.

Seeing Dad read the report card with skepticism and then pass it to my overjoyed Mom made that summer studying in the bedroom worth every minute, not to mention John's jaw dropping to the floor. More than a goal, straight A's had felt predictable, even inevitable.

All three fall grading reports were straight A's. And more importantly they put me at the top of the lower-academics class; so much so, teachers began requesting my help tutoring other challenged students.

Still, my success created a problem. For most subjects, both upper- and lower-academics classmates took the same test, but then the two groups were graded on separate curves. By mid-year grading before Christmas break, the teachers had a dilemma—my scores were driving the grading curve lower for other challenged students. This had to be resolved by removing my grades before figuring the lower-academic-group curve.

One of my teachers recommended a midterm transfer from lower to upper academic class. Though an objection arose that it had never been done at TAHS, strings were pulled for the transfer and for the rest of that year, my scores went to and stayed at the top of the upper-academics group. In fact, my chemistry and mathematics instructors added more extra-credit questions to pull everyone else's grades up.

And again, several teachers requested my tutoring, this time for the higher-achieving students who had difficulty mastering challenging skills like chemical equations and slide rule, a mathematical nightmare for many pre-1970, science students.

The pattern repeated in my senior year, spiriting off textbooks and doing all the problems ahead to prepare academically—even college-preparatory subjects like calculus. At graduation the class awarded me "Most Likely to Succeed" and teachers added honors medals in science and math.

The greatest rewards came from building relationships through tutoring students who had struggled. Classmates recorded their appreciation in the 1966 TAHS yearbook by serving up a few nicknames like Elvis, Slide Rule, and Einstein. For me the storyline of those years followed the high-school rocketry of John Hickam's 1999 film *October Sky*.

During that time a seed began to grow, stirred by radio programs and my being inquisitive about knowing God personally. Prior to marrying Dad, Mom voluntarily converted from Presbyterian to Catholic and attended mass regularly, often walking a mile to church. Dad went Sundays, drawn to sing in the all-men's choir and playing organ when needed. Both parents were accomplished musicians on piano and multiple instruments.

Having quit school at age fourteen to work in the coal mines, Dad saved enough during the depression to purchase a beautiful, baby-grand piano. After many years, the piano's veneer started to peel—just as Dad's superficial faith developed cracks.

Though a wonderful father and provider, like Job, a string of family tragedies and setbacks triggered Dad's anger. At times he would sit to watch (and usually rail against) TV evangelists. Some of the family reasoned, "Why not just turn it off?

He would angrily reply, "They make me so mad!" Once, while I was reading a Bible, Dad entered the bedroom and said, "That Book will drive you crazy!"

At that time, the life story of George Washington Carver and his personal faith provided me with inspiration during summer reading. Later in 1966, the *Revivaltime* radio program printed a widely distributed leaflet written by Dr. Wernher Von Braun titled, "The Farther We Probe into Space, the Greater My Faith."

During the Second World War, Dr. Von Braun had been the leader of the German rocket program at Peenemunde, and then moving to America after the war, he became head of NASA's George C. Marshall Space Flight Center.

Throughout my high school years, he was my inspiration in rocket propulsion. In a June-1966 publication, Wernher also professed personal faith in Christ, offering a moving testimony of the transforming power of the gospel message. At one point while viewing the night sky, I challenged the Lord, "God, reveal the meaning of space, time, and the universe."

Back in 1890, my great-grandfather William and his family landed at Ellis Island, New York, upon arriving from Scotland. They found residence in Western Pennsylvania near Johnstown, a city recently devastated and virtually wiped out by the Great Flood of 1889.

Prior to World War II and because of work in the coal mines, my great-grandfather William and grandfather John averaged moving every eighteen months.

Still in Pennsylvania, eventually the family settled in Ehrenfeld and finally Sidman near the old South Fork Dam (source of the 1889 Johnstown flood). Decades later with the local economy in an early-1960s tailspin, hope for desperately needed jobs glimmered when a large trucking company built a new depot in town. Regrettably, though, one night during a campaign to unionize the depot, the building succumbed to a devastating blaze. Rumors abounded along my newspaper route.

During that time in Western Pennsylvania, political tensions and labor unrest were unsettling. Scanning the newspaper before delivering my route, I saw photos of truck windows shot out as a warning. One featured the shattered windshield of a coal truck driver who had been hauling during a work stoppage.

Observing my disapproval with what was happening, Dad cautioned, "Don't ever cross labor leaders or politicians around here," and then added apologetically, "After a shift in the mine, I won't participate when they rally everyone to get their guns to take care of some business."

His father, John, worked the coal mines and assisted union organizing under John L. Lewis, founder of United Mine Workers of America (UMWA). Coal and union culture were rooted deep in our family. From decades of coal sulfur runoff, lifeless orange streams and rivers became characteristic of the area.

After years of being unemployed, Dad finally got back into the coal mines as a mechanic, some of which reopened as Japanese interests purchased abandoned mines to import fuel for overseas steel mills.

Long after having spent much of his Army days during World War II on Guam in the South Pacific—where America fought Japan—Dad found himself working for Japanese owners, a painful pill following years of personal, financial setbacks.

Eventually he tested for and received a position as a federal, mine-safety inspector for what was then called the Mine Enforcement and Safety Administration, ending Dad's long run as a coal miner.

In December of 1969, his warning held true when union president Anthony "Tony" Boyle hired hitmen to execute Jock Yablonski, as well as his wife and teenage daughter. Jock had challenged the election results for union president.

With steel mills closing, coal mines slowed, and with little but government welfare left, a dark pall descended over Western Pennsylvania. Common through Sidman until the late 1950s, steam-powered coal trains ended up being displaced by diesel locomotives. Earlier times saw seemingly endless lines of coal cars blocking railroad crossings, but by the late 1960s, train traffic had reduced to only a minor inconvenience.

Though slow-moving coal trains created delays on my newspaper route, they also served as handy transportation when hopping five-mile rides to visit friends in South Fork—with my 22-caliber hunting rifle.

Riding a Coal Train

Locally and nationally the landscapes were changing in troubling ways. Looking back, it seemed a portrait sketched out of Don McLean's "American Pie" lyrics, a scene lacking hope.

In later years while enrolled in Aerospace Engineering at Penn State, my hopeful sights for after graduation were set on Florida's space industry. The exhilarating success of the 1969 Apollo 11 moon landing became an isolated exception to the national mood.

With the Apollo program winding down, getting into that industry would become as challenging as my high school academic turnaround. But I learned that impossible doesn't exist. As it turned out, after completing undergraduate studies and enrolling for a Master's Degree program at Penn State through Air Force ROTC, my plans would be interrupted by a personal encounter with God.

Flight in a Lockheed T-33 Trainer Jet

During summer ROTC training camp each candidate had the opportunity to fly backseat in a Lockheed T-33 fighter-jet trainer at Myrtle Beach Air Force Base. But after boot camp, a routine physical revealed my eyesight did not qualify for pilot or navigator training, which put a restriction on Air Force career alternatives.

The options were to accept discharge or file for a waiver. I elected the former. Today after Lasik surgery both eyes are 20/20, an option not available in 1970. In retrospect, though, all these events built upon God's ultimate purpose.

Haiti Bible School Construction

Chapter Three - It's Broken

OVER THE COURSE of several years in the early 1970s, our local missionary team working with World Ministry Outreach (WMO) in Haiti varied from six to twelve of us. We managed to complete many outreaches in churches and several construction projects. Notable among the latter was the erection of a 30' x 100' cement block and concrete building for the purpose of a Bible school.

A major challenge involved forming and pouring the building's reinforced concrete roof. Calculations showed it would require thirty cubic yards of concrete, poured over two days. A second challenge was the lack of available lumber needed to form the complete concrete roof slab.

Virtually all lumber must be imported into Haiti due to defor-estation from charcoal's use in cooking. While working there, we found that any sturdy tree might disappear suddenly overnight, having been reduced into a smoking mound covered with soil for reduction into charcoal.

Occupied or not, wooded property might be stripped in a matter of hours to gratify the insatiable market for anything resembling charcoal. And with imported lumber subject to a 100% duty, purchasing the necessary wood for our projects became impractical.

A perfectly-timed blessing appeared when a nearby hotel underwent demolition for the construction of a new casino. The wood-frame construction of the demolished, 1930s-era casino lay in piles where the modern resort would be built.

When I explained our dilemma and offered to remove the nails, the manager graciously agreed to let us use the old wood. He provided a large quantity of beams on the condition they would be returned in time for his concrete-framing project. God always makes a way.

Lacking access to premixed-concrete equipment, the mixture would need to be prepared at the site, elevated ten feet to the roof, and moved across it by wheelbarrow. The job would be impractical for hand mixing, which meant renting a cubic-yard-capacity mixer and a lift that could handle one cubic yard of concrete—both machines gas-powered.

After some investigation, a rundown-but-functional mixer became available at an equipment yard. Unfortunately though, the only powered lift suffered the fate of most equipment in Haiti. In Creole, the owner said, "Li an' pan," meaning, "It's broken." That's also the *de facto* national motto of Haiti.

When asked how soon it could be repaired, the operator of the yard claimed he didn't have the money. Not willing to let a simple matter like that rest, I asked if—given the funds—he could have it ready for our scheduled concrete pour. He responded affirmatively and we sealed the deal, so with only minor setbacks, a few weeks later we completed the concrete roof.

One of the first projects in Haiti involved a Yamaha 250 trail bike. Before arriving, I had been informed a trail motorcycle would be available, so disappointment hit upon seeing the bike partially disassembled. Lacking a manual meant laying the motor out to determine the design and function of each part.

As it turned out, the only problem was a broken shifter yoke, which explained the disassembly. After sketching an image of the yoke and sending it back to the states, a replacement arrived and the trail bike came to life again. *Have a goal, make a plan to get there, circumvent every obstacle, and don't rest until complete.*

Yamaha 250 Trail Bike Repair

Similar complications with Land Rover and Jeep vehicles continued almost daily. At one point the Land Rover differential broke a tooth, resulting in stripped ring and pinion gears, despite my insistence it not be driven until repaired. In retrospect, the ignition key should have been hidden.

Unlike most stateside service shops, this dealer's service bay consisted of no more than a greasy pad in the back room. A young Haitian service technician got the ticket to overhaul the differential, and not being comfortable leaving the work to a stranger, I stayed to monitor the process.

Land Rover Differential Repair

Though the young man appeared to know the steps for removal, replacement, and reassembly, he curiously performed obscure practices. For example, rather than measure the gear backlash check with a dial or feeler gauge, he used a paper match rotated between gear teeth to set the backlash position. If the match crushed, it was too tight.

When asked if he understood the operation of the differential, he replied, "No, but I can fix it." As it turned out, the mechanic himself received no pay as apprentice. The dealer allowed him to work and learn the trade for a time until he became skilled enough to receive some form of certification. Then pending approval, he would have a job.

City water for our mission facility flowed only six hours each day, requiring storage in a large tank on the concrete roof. An electric water pump filled the four-hundred-gallon tank, which resulted in a big inconvenience when the pump failed. In order to replace it, the power and water lines were disconnected.

Not realizing it was a 208-volt system with *three* hot leads and a ground, I stood in the pool of water to lift the heavy pump from its mounting bracket. What happened next were shooting, internal fireworks that make me drop the pump like a hot Haitian sweet potato. For future reference I noted that the 208-volt system has three hot wires, and I thanked God that He kept me around to use it.

Not long after that shock, a circular saw got acquainted with my left index finger, cutting it to the bone. I knew something was wrong when noticing the streak of blood on the wall. My coworker David rounded up pliers, needle, and thread to stitch my hand back together, but it took a few years for the finger's nerves to be coaxed back into service.

Over many decades of foreign ministry, Haiti garnered the dubious title "Graveyard of Missions," for reasons clearly evident after a few years of work there. Frequently, new visitors there for a short period would display well-intentioned but misplaced priorities.

One guest I saw got a reality check upon gifting a soccer ball to a young boy who had been playing with a wadded-up ball of rags. His shiny new ball was immediately confiscated by an older boy. Another time a well-meaning coworker gave money to a girl on the street, only to see her instantly attacked and her windfall stolen—no good deed went unpunished.

David Skinner Preaching on a Jeep in Haiti

Social, financial, and governmental deterioration has accelerated in Haiti over the past fifty-two years since I worked there, eventually degenerating into gang warfare. Though it was safer in the early 1970s, risks still abounded. While contracting a carpenter to do construction of the school building, he requested and received several advances on his pay. After completing only a sixth of the work, he demanded more money, and when negotiations deteriorated, he showed up with the local Tonton Macoute.

Years prior under the leadership of former President Papa Doc Duvalier, neighborhood thugs were recruited to protect Duvalier against military uprisings. Each paramilitary Tonton Macoute received a hat, blue-denim uniform, and a vintage 1903 Springfield bolt-action rifle. Most Macoute maintained little discipline on use of authority or rifle. The denim-clad Macoute (Bogeyman), as they were "affectionately" named, arrived drunk and with rifle in hand.

After a brief exchange, the Macoute fumbled through the routine of loading his rifle and reinforcing his demands by dangerously waving the barrel in my direction. Our ministry's young ladies began voicing a chorus of advice, "Just give him the money!"

But after getting nowhere with me, the contractor and Macoute threatened to go to a local judge, which I knew would have assured injustice. Judges there pocketed most of what they managed to extort. I took the safer route—and the scripturally advised one—meeting the carpenter's demands and then discharging him.

Facing a Haitian Tonton Macoute

Similar run-ins seemed routine. During an organized tree planting near the Port-au-Prince Airport, our whole group was detained by a band of Haitian-military officers. Though the complaint said we had entered restricted military space, the misunderstanding got cleared up and all were released without charges.

Another time, a serious political crisis developed when a Haitian army contingent decided to overthrow then President Jean Claude Duvalier. The early-morning air resonated with cannon fire, and bursts of smoke could be seen across the bay near the Presidential Palace.

No revolution deserves to pass without investigation, so I headed the Land Rover in the direction of the Palace where the uprising appeared to dissipate as quickly as it had started. A group of Christians lay prostrate on the street, crying and praying for the nation and perpetrators, who were reportedly disbanded and dispatched—like all Haitian revolutionaries.

Despite challenges, all things worked toward positive outcomes. The ministry employed up to twenty locals for operation of the school, clinic, church, orphanage, and construction. Once when our funding didn't arrive for three months, Haitian employees and neighbors stepped up with food and other resources.

We ate regularly despite some adjustments, like the problem that needed to be solved with several fifty-pound bags of bulgar wheat donated the year before from the USA CARE agency. Our team refused to eat the grain because it had become contaminated with meal worms.

Without notifying anyone, I loaded the bags on the Land Rover and transported them to a local processor to have them milled into flour, reasoning that everything in the worms then was first wheat. Upon my return, the team marveled at this miraculous provision of ground farina. No further complaints were heard.

When a woman visiting the ministry took a few orphans to the Iron Market at the center of Port-au-Prince, the children clamored for some canaries displayed in small wicker cages. Not anticipating the outcome, she purchased each child a pet canary.

But back at home later, we noticed some orphans walking out of the kitchen crunching on a roasted bird. Native birds weren't safe either. Any bird perched on a power line often fell prey to a local's deadly accurate stone throw.

And rumor has it Haitian cats quickly deplete all nine lives. At one point when Debby decided to get a cat, she found a kitten and allowed it free run of the house. When it disappeared one day, Debby inquired of the neighbor, who replied, "I'll get it back for you," which they did.

A few days later it must have managed to escape the house again, but this time the neighbor could only apologize. "Sorry," he said, "they already ate it." For those claiming Haitian cat meals are unsubstantiated rumors, you had to be there.

We regularly faced the cultural challenge of voodoo practices. Occasionally a church-goer would come forward for prayer, wearing a fetish around their neck (an object supposedly possessing magical powers).

Questioned about the fetish, the person would usually reply, "I bought it from the ougan (witch doctor) to ward off spirits." Of course, driving away demonic spirits meant that item had to go first. In another instance a neighbor explained that he needed to pay the ougan to chase spirits from his field—for fear that "someone nearby might burn my field to prevent the spirits from coming over to ruin their own crops." Every fear has its fee.

Lest these events be interpreted as a picture of disappointment, I'll relate how they all fit into God's broader outcome of positive change. Many of the Haitians and some missionaries involved during our time still work in ministry fifty years later.

Twenty years after our move away, Debby and I returned to the same neighborhood in the Arcachon district of Carrefour to find that our efforts expended decades earlier contributed to the creation of a flourishing neighborhood, a still-growing church, and voodoo practice all but eliminated.

Even after much destruction with the 2010 earthquake, the time our ministry had been able to spend there continued changing lives and even impacting nations beyond Haiti. Our coworkers Joel and Yvonne Trimble have a legacy of working more than fifty years in their ministry, Haiti for Christ, as well as their La Bonne Nouvelle broadcast that reaches Haitians with TV and other video from around the world.

Jim and Vicky Gross minister with Teen Challenge, and David and Maria Skinner operate Network Bible Translators, furnishing Bible translations to remote tribes.

Today, after having worked design and installation of gas-turbine-driven power plants in many nations of the world, remembering Haitian conditions always provides a reality check for the way much of the world lives. Many nations and cultures around the planet fail to demonstrate resilience and resourcefulness to sustain their lifestyle without outside assistance.

A recent TV broadcast featured several third-world individuals being asked whether their languages had a word for *maintenance*. Each replied no to the apparently foreign concept. To follow biblical principles, the Church must be capable of functioning internally wherever they are *and* reaching others beyond.

Outreach is essential for spreading the gospel, with the express intent of empowering others to go and do the same. A necessary aspect of discipleship involves reproducing yourself to enable others to reproduce themselves, and so on. Disciples make disciples who make disciples.

Rather than surveying the world, declaring it's broken, and giving up, that problem can become part of the solution. After decades making furniture, I've found one constant is ending up with a flaw. It can happen from absent-minded repetition, a wrong measurement, or a conspicuous mistake.

When a flaw is found, the first course of action is asking, "Can it be fixed?" Next, "How do I fix it?" And lastly, "Is it worth the cost?" In the final analysis, "Repair or replace?". But leaving it broken should not be an option.

In God's economy the same applies to the world. The Lord allows resistance in the form of evil that provides training—a "workout" in a "fitness center" that prepares us to work against the resistance. Cynics complaining how a good God could allow evil in the world sounds surprisingly like a visitor to the fitness center criticizing the presence of heavy weights. Removing heavy equipment certainly seems to make the exercise room a safer space. But as King Solomon observed,

Where no oxen are, the crib is clean: but much increase is by the strength of the ox. (Proverbs 14:4)

Jesus did *not* come into the world to talk about the problem, declaring "It's broken" and then departing on the other side. Rather, He came to do something toward fixing the issue, having us work to overcome the resistance—instead of God just eliminating the world's evil.

During his Olivet discourse, Jesus forewarned His disciples of events to come, telling them the parable of the shepherd who separated sheep and goats (Matthew 25:31-46). Those who reach out to others in this life will be invited in by Christ because they saw needs and invested in others. Whereas the goats were excluded because they closed their eyes and ignored the need.

…Verily I say unto you, inasmuch as ye have done it unto one of the least of these my brethren, ye have done it unto me. (Matthew 25:40)

Gathering Manna in the Wilderness

Chapter Four - What is It?

WHILE LOOKING FOR a church after moving to Greenwood, Indiana, we were encouraged to check out a group assembling in an empty store front downtown. Having just relocated for a new job and with a home under construction, we hadn't expected to step into a church-building project.

At the time, about thirty couples and singles started as a home church under the leadership of Dr. Charles Lake and his wife Vicki to create Community Church of Greenwood (CCG). Prior to that beginning, finding time for personal devotions and other spiritual disciplines had become increasingly sporadic. Foundational to the concept of CCG was a series of four, nine-week courses with the Discipleship Training Program (DTP) developed by Dr. Lake.

Central to the program, participants had to maintain a daily devotional time of Scripture reading and personal application. Other subjects encompassed a spectrum of disciplines including prayer, memorization, fasting, witnessing, tithing, and the list goes on. That program provided a foundation for explosive growth in the church and CCG membership.

After completion of all four courses, the opportunity came to teach several sessions, both as a class and one-on-one. After our move to the Cincinnati, Ohio area, the discipleship program and model for ministry were incorporated in other churches.

Recognition of the importance of personal daily Bible reading, application, and prayer establishes a foundation for spiritual growth and provides opportunity to biblically influence others. Daily intake of God's Word and personally applying it supplies each day's foundation. Upon establishing that, everything else comes into proper perspective. Jesus portrayed Himself as the daily provision, saying, "I am that bread of life" (John 6:48).

Find a consistent time and place for your devotions. Keys to effective discipline at prayer and scripture reading can involve tools such as some version of the *One Year Bible,* which facilitates daily reading and personal application (while completing the entire Bible in 365 days). Tools for focusing on the purpose for prayer can include personal needs, prayer lists, and the *Operation World* daily prayer book or phone app that covers prayer for all nations of the world over the course of one year.

Through the period I was involved with discipling, Steve (not his real name) stands out. He asked to participate in one-on-one training through the nine-week Discipleship-1 course. After a couple of weeks, he began showing up late or not at all, and rescheduling no-shows created conflict with other evening training sessions.

After the situation deteriorated further, we mutually agreed to suspend his classes. A couple of years later while living in Cincinnati, we visited family near Greenwood. An acquaintance informed us that Steve had finally completed the training and was effectively influencing others. Even in failure, God ultimately finishes His work and gets the credit.

Fritz Thomas with Haitian Neighbor Boys

Another example of discipleship came to light out of our work in Haiti. During the years WMO maintained an outreach center in Haiti, Fritz Thomas persisted as a neighborhood holdout. As we walked Arcachon Road along Carrefour, Fritz frequently stood alongside the road surrounded by a group of mid to late teen boys. Tall and muscular, Fritz highlighted his presence by parading without a shirt. On the way down the street to start church service, I often called to Fritz, "Come to church service with us." His response was usually, "You go and start the service. I will join you shortly." That was typically followed by a chorus of laughter from his court of admirers. They were certain he would not be seen in church. Recently I had the opportunity to visit Fritz and Ghislaine Thomas at their fiftieth wedding anniversary. Ghislaine had lived with us in our orphanage along Arcachon Road. Jean Fritzner Thomas now pastors Eglise de Dieu Haitienne in Port Charlotte, Florida, a short distance away from our home. Despite our best effort, God always completes his purpose in his time, because a seed always bears fruit in its' time.

The daily bread symbolism of discipleship training reminds me of some food mentioned in the Bible. Upon exiting Egypt, the traveling Israelites quickly learned the importance of a stable nutrition source. The wilderness they went through offered no convenient meals, or at least not their typical food.

Under most circumstances a trip on foot from Egypt to Canaan (Palestine) should have been an eleven-day walk in the park, compared to the forty years it took them. A similar trip mentioned in Deuteronomy 1:2 from Horeb to Kadesh Barnea (Egypt to Palestine) also tell us it takes eleven days.

At that time food and water were transported by satchel on a donkey, with the owners riding the animal or walking beside them. But given the logistical challenge of moving almost two million people out of Egypt, along with their herds and belongings, it took the Israelites longer than eleven days—though it shouldn't have taken forty years.

Told to leave Egypt by God, once the travelers realized the challenge of getting food and water, complaints began. In response to the murmuring, and besides occasional quail, God supplied manna for their food through the entire journey to Palestine (Exodus 16:4-36).

What is manna? The meaning derives from the question that means, "What is it?" The people responded to the unfamiliar food by asking what it was. This was prior to Amazon air-dropping deliveries in the Sinai desert, so God provided a unique solution in manna.

What was the makeup of manna? While the composition cannot be known, it can be said that manna contained all the nutritional value needed: vitamins, minerals, carbohydrates, fiber, and protein. The appearance of manna resembled coriander seed and tasted like wafers sweetened with honey.

But a problem developed as some of the people began to consider the same food each day to be monotonous and started complaining about the uniform diet (Numbers 11:4). They desired flesh to eat so the Lord again accommodated with quail, but as soon as it was in their mouths, the Lord struck them down at a place that was named Kibroth-hattaavah, which means "graves of lust."

When questioned about what miracle He could perform on par with Moses supplying bread in the wilderness (John 6:30-35), Jesus proclaimed, "I am the bread of life: he that cometh to me shall never hunger; and he that believeth on me shall never thirst (John 6:35)."

Manna's mystery was how it supplied everything necessary for life. Similarly, the Bible's shewbread (face bread) gave the people face time with God (it was Him in their midst), and Jesus spoke metaphorically about the bread of life, "Man shall not live by bread alone, but by every word that proceedeth out of the mouth of God" (Matthew 4:4).

The importance of gathering manna daily was a takeaway from my experience with discipleship training. As with that wilderness bread, the experience of discipleship cannot be described by anything familiar because the bread of life or Word of God develops inside a person as they have that daily relationship with the Son of God.

And like manna, it's best to begin the day with it rather than wait for the sun to come up. Establishing that discipline first thing in the day brings a foundation that helps all else come into focus.

Since that initial experience in Greenwood, Indiana, teaching and training others has been a lifelong pursuit. And discipleship isn't about being a better Christian; rather, as a follower, it's listening to God's command to influence others: a disciple makes disciples who make disciples...

Speaking of the shewbread, the Lord instructed Moses to make a table overlaid with gold, dishes, and spoons of gold (Exodus 25:29-30). On this table Moses was to set bread before the Lord (or shewbread).

In Hebrew, the expression for "shewbread" means "face bread."

Every aspect of the Lord's tabernacle furniture and function expressed relationship with God.

The contemporary expression mirroring "face bread" would be someone getting "face time" with God. In maintaining a meaningful relationship with the Lord, face time prevents the traps of isolation and self-reliant thinking. For that reason, the priest renewed the table regularly with fresh bread. The shewbread principle intertwines with the provision of manna in the wilderness.

Just as manna sustained the Jews during the forty-year wilderness trek, today God supports His believers with everything needed. Whether now or in Biblical times, the availability of God's Word or manna necessitates people waking early to have their devotions or to scour the plains around the camp to gather food for the day.

Manna did not last into the next day (except on the Sabbath), and in our time in God's Word and a relationship with Him through Jesus Christ (The Word of God) He supplies everything we need for the day, when obtained early.

Just as the Israelites witnessed God's sustaining help every day for forty years, all the way to the Jordan River on the boundary of their Promised Land in Canaan, so also, we must stay in the Word daily—as our day of rest and final redemption approaches when we'll be in Christ's presence—where there's no need for symbolic bread.

F101-100 Jet Engine Test

Chapter Five - Fasten Seatbelts

B Y THE MID-1970S, the Arab Oil Embargo heavily impacted engineering job opportunities in Florida. Sensing the calling to continue in the engineering field after leaving WMO, Debby and I moved north in 1975 to Indianapolis, Indiana, and then Cincinnati, Ohio. In each location we worked and ministered with existing churches or helped plant new ones.

During that time while working at an equipment manufacturing plant in Indianapolis, a customer placed an order for design and fabrication of an assembly-line furniture dryer. After laying out the requirements and completing a design, it was fabricated and shipped to the customer. It was then that I realized a section of the dryer did not have sufficient flow area to provide the required heating.

Reluctant to publicize the error, only a miracle could avoid a major blow-up between the business owner and this regular customer. As expected, a few days later the owner came into the office announcing the customer received the shipment and planned to return the dryer for changes.

It turned out the customer had provided the wrong width dimension for the assembly line and would cover all the costs of modifications. Upon making corrections for width and flow area, the dryer again shipped out and functioned as intended. Count that as another miracle.

In early 1976, an aircraft-engine manufacturer in Cincinnati offered me a position in aerospace engineering for jet-engine development. That opportunity opened prospects to work on systems-integration design for cooling and lubrication on numerous commercial and military jet engines. Most notably that included the joint venture with a French company on various versions of the CFM56 engine, and for F110-100 and F110-400 fighter engines.

My responsibilities included the anti-ice system for a then-secret F118 engine on the future B2 stealth bomber. Later roles included developing aeroderivative gas turbines for power generation. In the process, a total of ten patents were issued, and numerous technical papers and engineering manuals published. Over thirty-six years in aviation and industrial, gas-turbine-engine development opened doors to visit many customers in countries and cultures around the world.

As part of our company and the University of Cincinnati Master's Degree program, advanced engineering courses were provided at the manufacturing plant. Debby jokingly declared she had thereby obtained her M.R.S. in Aerospace Engineering. After completing the advanced course graduation ceremony onsite and along with several other couples, Debby and I toured the jet engine test cells.

In one of those an F101-100 engine for the B1-B bomber ran through cyclic testing at full afterburner. As we approached—and despite thick walls—the floor and ceiling shook. Against a deafening roar, the engine appeared through the viewing window with a long tail of shock diamonds highlighting a blue, supersonic exhaust. I looked at Debby and said, "Isn't that amazing!" She glanced over at me and replied, "Sure. Where are we going for lunch?"

One summer a challenging problem developed when assembly personnel were out on strike and the engineering manager asked for his design of a new gearbox to be assembled. The parts had just been delivered for the CFM56-2 engine inlet gearbox to be installed on a special YC-15 STOL (Short Takeoff and Landing) demonstrator aircraft. I agreed to complete the work because the gearbox delivery required a tight delivery schedule.

Using a preliminary set of assembly instructions and a box of parts, that gearbox and two others were assembled in a few days and all performed successfully. However, after the work stoppage the returning assembly technicians expressed displeasure with the assembly work having been finished without them. It took time to rebuild bridges with the team.

Shortly after joining the company, several coworkers invited me to join a noon Bible study on Wednesdays in one of the conference rooms. At that time many of the company buildings had a noon Bible study group. Rather than just a footnote on the job, in times of difficulty those groups became essential to the culture and as an influence at the company. Through over thirty-five years participating in weekly Bible study meetings, for at least twenty of them I was able to teach biblical subjects and topical series.

My career at the company began under the leadership of Gerhard Neumann (1917-1997), a Jewish German engineer who worked in Hong Kong prior to World War II. Gerhard found himself with no means to return to Germany as Japan attacked China.

Desperate to make alliances, he connected with Claire Lee Chennault, who led the American Flying Tigers in China. During the war Neumann rebuilt a wrecked Japanese Zero to help American pilots understand the plane's capabilities.

After the war during a takeover by the Communist Chinese, he and his wife Clarice drove a jeep out of China through Asia and the Middle East to Tel Aviv, Israel.

Upon landing a position at the company, Neumann's invention of the compressor variable stator vane propelled him to the executive leadership role in jet engine development. He also founded the international organization whereby the company and a French jet engine manufacturer developed the CFM56 series of gas turbine engines (where my career started).

In recognition of his efforts, Gerhard Neumann was the first person granted US citizenship by an act of Congress. His autobiography, *Hermann the German* (1984), captured his story and career. Late in his tenure at the company he had been hospitalized with a serious illness, so our Bible study group sent a signed card to his headquarters office on the floor above our CFM56 engineering area, notifying him that we were praying for his recovery.

Upon release from the hospital and his return to work, Gerhard Neumann made a point to come to the Wednesday study and offer appreciation for the concern and prayer. Attendance at the studies varied from just a handful of people to the occasionally full conference room. After one of the lessons, a consistent attender approached me to confide, "When I began participating in the Bible Study, I was the worst drug abuser in this place. But now my life has been changed." Over the decades, several participants left the business to enter pastoral ministry or missions service.

Generally, a topical study series—like end times, evolution versus creation, or Christian apologetics (defending the faith)—meant posting bulletin-board fliers to spark interest.

On one occasion a senior section leader notified my manager that Bible studies must stop because it wasn't allowed in the building. After I drafted a protest letter, insisting that stopping our lunch gathering meant ending all non-business meetings at lunchtime. The section leader consulted with legal and human resources departments and quietly withdrew the demand.

In another instance, the head of our business group tore down several of the Bible study fliers. Later when he was going through a time of personal difficulty, our Bible study participants lifted him up in prayer.

His secretary was an outspoken witness for Christ, and she mentioned that he committed his life to Christ before passing unexpectedly. Each confrontation and challenge resulted in God's supernatural provision for resolution.

CFM56-3 Engine and 737-300 Aircraft

As part of the CFM56-3 engine design and development, our testing of compressor bleed air for cooling resulted in pressures inconsistent with prior, published results. During the test program, I made a presentation to engineering management, claiming the latest measurements must have been incorrect and so I was electing to stick with the early data.

Within a week, while reviewing prior analysis reports, it became apparent that another colleague had created the earlier "data," based on a hand calculation. From there the alternatives were admitting to leadership that my prior assessment was mistaken or just keeping it to myself.

A favorable outcome happened when I went to the review team and explained how the early results were not test data but calculated. The resulting redesign opened the door to an improved compressor-bleed system that dramatically improved flow efficiency. Reporting the error along with fixing the problem resulted in a significant design patent award.

During a series of gas turbine tests for CFM56-3 engine seals, results were presented at a review meeting with senior leaders. One recently promoted technical leader made a statement inconsistent with results collected from one of the tests.

Upon my pointing out the latest results, he glared, taking offense at what he seemingly considered a personal affront. The next day, upon entering his office prior to a scheduled meeting, he quietly assured, "I am going to get you fired from this business."

Though I took this as a jest, the threat turned out to be starkly real, and later he came to my office, demanding all my documentation for the work completed on engine-cooling projects. Over the next several years, he made repeated efforts to investigate my work assignments, looking for justification to encourage dismissal.

At another time after the same leader made an impassioned plea to senior engineering management about firing me, my personal manager said he understood the situation and was running interference for me with company leadership. God ultimately maintains control of all things.

The Bible study group provided much support during that period. Ironically, after that technical leader passed away some years later when I was in a senior technical position myself, our team was assigned to remove several unnecessary product changes that the technical leader had implemented. They were not justified by effectiveness and cost. A Scripture encouragement during that time was God's affirmation to Joshua, "There shall not any man be able to stand before thee all the days of thy life… (Joshua 1:5)" God works His mysterious ways. No hard feelings.

Years later in the mid-1990s, Debby and I applied for and were approved for mission service again. However, due to our age the stipulation from the missions' board required us to be self-supporting. In other words, we would need to use our own resources rather than officially raising support.

That was not practical at the time, so we continued to work while ministering through churches. During that period, development of a teaching series on end times continued with the workplace Bible study, as well as in church training lessons.

After many years working on design and development of aircraft gas turbine engines, the opportunity came to work in the design team for the LM6000 Aeroderivative Gas Turbine. The project provided immense opportunities worldwide for power-plant projects. With the completion of design and packaging, each installation required a pre-start Operational Readiness Review (ORR).

LM6000 Aeroderivative Gas Turbine Engine

One of the early customers purchased a twin-engine power plant on Hainan Island, part of Mainland China. In July 1995, our team travelled to China for the on-site review and to finalize plans for startup and commissioning.

Having frequently used Jack Chick cartoon tracts for witnessing, several hundred Chinese-language versions were purchased to hand out in the villages near the power plant. Some attendees at the Bible study cautioned the tracts risked confiscation in Chinese customs.

Decades before when carrying Bibles into Russia, the founder of Open Doors ministry, Brother Andrew, had prayed for God to make seeing eyes blind. After pursuing this tactic myself, I watched our

Chinese customs official open the suitcase, give a quick look, and close it. I walked out of customs with relief.

After that, our team assembled on a balcony overlooking the airfield to await our next flight to Hainan Island. Passing by, a group of young locals conspicuously announced our arrival, yelling "CIA!"

When we spotted a line of Chinese, MIG fighter jets on the runway apron, a colleague absentmindedly pulled out his camera to take a picture. I yelled at him to stop, hoping none of the security guards noticed, and instantly he realized the implications. We could have been detained, jailed, and deported. Again, by God's grace no one noticed.

Over the next few days, the review for startup was completed successfully, and after handing tracts out on a number of walks through the region near the power plant, residents could be seen passing them around.

Though only small seeds were sown during that visit, decades later there were published reports of a great revival sweeping the region of Hainan Island and the larger province. The catalyst had been much more than the tracts, as many evangelists from China witnessed in the region.

Ultimately, God completed the work, brought forth results, and He rightfully receives the credit. Chairman Mao Zedong achieved the unlikely title of "World's Greatest Evangelist." This followed after his mission of resistance to organized religion. In the wake of his religious purge, Christ's true gospel prospered. Also, the gas turbine generators started and operated successfully.

LM6000 Inlet Filter Fire Investigation

During my role leading the Aeroderivative Turbine Installation and Service Safety Team, potential safety concerns were addressed as well as failure events that required investigation and reporting. A year after qualification and startup of an LM6000 power plant in Spain, the customer reported a fire in the inlet filter house. The filter system looked like a total loss.

Each power-generation package includes a very large, HEPA, filter-house assembly. Back then the filter-house cost was nearly a million dollars. From first communication, the operator claimed there must have been an electrical short in the inlet filter assembly and insisted the company cover replacement.

This installation served as a power and steam generation for processing olive oil from a nearby orchard. Olive pits burned in a furnace provided additional heat for the steam process. Not surprising, the machine's filter-canister elements were saturated with olive oil mist, a byproduct of processing.

After an extensive review of findings, no conclusive root cause of the fire could be identified. In such cases, it is customary to list most likely contributors and suggest one or two as the most-probable cause(s).

Potential sources of the fire were electrical short, smoking, lightning, sparks from ash, and static electricity. Writing the event report, ash sparks and static discharge were identified as the most-probable causes. As expected, the customer took exception and focused on a potential electrical short, despite lack of evidence.

After offering a discount, the inlet filter house was repaired and replaced. As always, in life and in root-cause investigation, I argue that "a root remaining is a cause in training." Within a year, the filter house caught fire again, and in this case it happened during a windstorm where embers from the ash bin had been observed blowing into the inlet filter. All things eventually come to light, as duly noted in the second failure event report.

One of my employees in the cooling-system-design group, Jim LeRoy, had a dream of becoming a stunt pilot. After leaving the company, Jim entered training, eventually becoming the world-class, stunt-pilot Jim "Bulldog" LeRoy.

He called me in the summer of 2002 to announce he would be performing at the Dayton Vectren Air Show. The demonstration required two assistants to hold the poles for his ribbon cutting stunt. My son-in-law Tim and I were able to reconnect with Jim and partic-ipate in that event.

Five years later in 2007, he was tragically killed during a presen-tation at the Dayton Air Show. At his wife's invitation, Debby and

I attended the funeral in Chicago that year, where an outstanding evangelical message was delivered about Jim and his influence.

Stunt Pilot Jim "Bulldog" LeRoy

Over the course of fifty years in the career fields of aviation, power, and space, the total product and service sales value I worked on amounted to well over a trillion dollars. Today my small contribution to each of these products is working around the world, creating trillions of dollars in value and influencing the lives of all eight billion people.

But that pales in comparison to the lives Debby and I have been blessed to touch and change through witness and teaching of the gospel. As we see in His plan, God reserves the glory unto Himself for His completed work.

A Tale of Two Men

The principle of God's purpose stands out in the story contrasting two men during the reign of King Hezekiah (715-686 BCE). The prophet Isaiah introduced two men with uniquely different outcomes (Isaiah 22:15-25). Scripture offers little background about Shebna (meaning "growth") other than he was King Hezekiah's treasurer and chief of staff.

Shebna occupied a prominent position in the government. But it seems he inflated himself by self-promotion, as when he carved grand sepulcher out of rock. Isaiah prophetically denounced Shebna for self-aggrandizement, saying he would be cast out and another take his place.

King Hezekiah, Eliakim, and Shebna

Isaiah pronounced the second man, Eliakim (meaning "God raises"), son of Hilkiah, would become the replacement for Shebna. This man received the robe of Shebna and his position, acquiring government into his hands and becoming the father to Jerusalem.

In a series of elegant metaphors, God promised Eliakim the key to the house of David that would be laid upon his shoulder. By that key, whatever he opened, none would be able to close, and what he closed none could open.

And God said he would be for a glorious throne to his father's house, which means, upon Eliakim, the Lord placed all the glory of

His Father's house, including all vessels small and great. In fact, he is represented as a nail fastened in a sure place. But then, the Word of the Lord says that nail was removed and the burden on it cut off.

Does that seem contradictory? Well, not if you understand context. From the meaning of his name Eliakim ("resurrection"), to the figure of the key of David (Revelation 3:7), to the burden on the nail removed, each metaphor points to this man as a picture of Jesus Christ.

Just as happened to Satan the usurper, Shebna is removed to make way for the Lord himself. Later in the book of Isaiah (Isaiah 36:11), Eliakim and Shebna appear again, making a stand against Rabshakeh, emissary of the King of Assyria. At this time roles are reversed with Eliakim being Hezekiah's chief of staff and Shebna only a scribe.

Though on the surface this appears simply to be a story of two men, their careers and organizational dynamics, the themes threaded through the Scriptures show how the person and work of Jesus Christ interweave the Word of God.

Eliakim and Shebna reflects God's sovereign control of personal destiny. In the same way for us today, as tribulation, deliverance, and the Day of the Lord approaches, Jesus Christ will quickly appear with signs in the heavens to catch away faithful believers and complete his role as Savior, King, Judge, Priest, and Lord of all.

Shroud of Turin

During the time I worked on the development team of the LM6000 industrial gas turbine, a role developed to work with multiple engine packagers for power-generation plants. On one of several business trips in the mid-1990s to a power-generation packager in Turin (or Torino, Italy), our customer engineers mentioned that the relic referred to as the Shroud of Turin had been placed on display for a limited time. Though our team drove past the line waiting to see it in a chapel, business commitments regrettably did not permit us the opportunity to stop and visit.

During a 1980s Bible study we had reviewed evidence for the shroud as the burial cloth of Jesus. Tests performed in 1978 by STURP

(Shroud of Turin Research Project) using image analysis and adhesive tape transfer resulted in the team making conflicting conclusions about the origin of the cloth image. Some researchers claimed there was evidence of pigments and no blood, while others disagreed on the basis of various tests or staked a claim of inconclusive origin.

Image on the Shroud of Turin

Then, in 1988, radiocarbon testing performed on several samples resulted in a collective dating of 1260-1390 CE (Christian Era) for the cloth's origin, implying a medieval-era forgery. With the publication of results at a news conference on the Shroud of Turin investigation, most of the media and voices of specialists breathed a collective sigh of relief, agreeing the matter had been settled.

Later at our weekly Bible study, I claimed, "They got it wrong." Even to the present day, many published technical assessments of shroud evidence reveal most authors and experts reject first-century, middle-east origin and provide their detailed, dismissive assessments of any data to the contrary.

That is still the case, despite 2024 Wide Angle X-ray Scattering (WAXS) tests that date the Shroud of Turin to the time of Jesus and even show scarce type-AB, male blood that is common to middle-east Jews.

Pollen from the area of Palestine was found—also supporting a first-century date. Without digging into the technical details of the evidence, how did I know in 1988 the STURP researchers got it wrong? Even to the present-day, researchers go to great lengths creating seemly plausible explanations and simulations to show how the image on the shroud "could have" been forged.

Similarly with a gas-turbine-failure event and determining the right, root cause through forensic investigation, a few underlying principles are observed:

1. Don't jump to seemingly obvious conclusion (more about that later).

2. The principle of efficient causation is sometimes referred to as Occam's Razor and it says this: When judging between conflicting explanations, the simplest one is usually best. While not a rigorous test, it provides a powerful sanity check to ward off complicated or sophisticated, technical arguments. Proposed by philosopher William of Ockham, the principle declared, "Entities must not be multiplied beyond necessity." This principle can also be referred to as piling-on. Occam's Razor also cuts through many of the hypothetical plausibility arguments of evolution and the investigation of the 1986 Shuttle Challenger failure as critiqued by Dr. Richard Feynman (again, more on that later).

3. Finally, human factors of investigation present a constant challenge. Interpretation of results can be clouded by faulty preconceived notions and assumptions, through which objective evaluation of evidence gets distorted—as in the case of that inlet-filter-fire investigation.

Likewise, the vision of most people is clouded about the soon coming of Jesus. First Corinthians 13:12 says, "For now we see through a glass, darkly..." But as the end times we're living in unfold for those paying attention, the season of Christ's return will come clearly into focus. (though the Bible tells us that no one will know the day and hour).

Artemis Astronaut Application

In June 2008 an ad appeared in *NASA Tech Briefs* magazine with a website link for astronaut applications. The stated purpose of the astronaut role would be a group of trainees in support of the upcoming Artemis program, using the proposed Space Launch System (SLS).

By that time, the Space Shuttle had been slated for retirement within several years. Reviewing the website and all qualifications, there did not appear to be any age limitation, though at my age of sixty, ten years of training was a long wait.

The application would be a long shot, but worth the effort—if only for bragging rights. Upon collecting all of the requisite documentation, I submitted the application by the deadline. After some back and forth communication with NASA, notification arrived that the application was accepted. The downside was the other 33,000 applicants.

For almost a year there was no follow-up from NASA until 2009 when a letter announced that only nine finalists had been selected, out of 3500 qualified applicants. I was not one of them. Rumors were that one of the missions would be a one-way trip to Mars. I joked that Debby was disappointed I wouldn't be going.

At my company retirement dinner five years later in 2014, my manager played a fictitious audio recording, supposedly from the former NASA administrator notifying I had not been selected. He also gifted me with an astronaut jump suit.

Then another five years later while I was doing contract work on rocket engine development in Florida, NASA funded a project for me to create ten training presentations on maintenance tasks for the RL10 EUS (Exploration Upper Stage) rocket engines—my consolation prize for not being allowed to visit Mars.

Artemis SLS RL10 EUS Rocket Engines

In all these unique circumstances, a recurring theme demonstrates how God is always working His plan for His purpose, against all forms of obstruction. And in some form or other, everyone experiences those issues as our resistance training brought on by persecution from our common antagonist, Satan, the arch enemy of God's will. More about that later.

Going back to 1992, a Bible study series on end-times events received significant participation and interest. Based on the study series, a foundation of teaching developed around the biblical stories regarding Christ's return.

This and other studies were adopted for Sunday School classes and small-group teaching. The past thirty years many participants encouraged publication of the information. This book's title, theme, and the remaining chapters of personal experiences are a compilation of those lessons.

During decades working in Indiana, Ohio, and Florida, Debby and I had the opportunity to assist in the startup of several denominational and independent churches, as well as teaching discipleship and serving on six church governing boards.

These efforts included jail and inner-city-Cincinnati ministries, along with tutoring students in an inner-city high school. At the time of the events surrounding 911 in New York City, news outlets warned of stock markets crashing at the opening on the following Monday. During a time of prayer for direction about our investments an answer came, "All your investments are safe and earning dividends."

That word did not come from within, and the realization hit that God wasn't speaking of the financial markets.

Why do the Peanut M&Ms float inside the ISS rather than fall to the floor?

If you say, 'There is no gravity', you are completely wrong!

Astronaut in ISS Eating Peanut M&Ms

Chapter Six - Space Station Window

PICTURE YOURSELF FLOATING inside the International Space Station (ISS), casually eating peanut M&Ms that drift around the cabin. In the background earth's curvature arches across a window. Why do the M&Ms float in the ISS rather than falling to the floor?

Having presented this scenario often, the most common response is that there is no gravity, which seems obvious on the surface. In fact, when describing this scene during engineer-training sessions, many technical specialists, scientists, and engineers also jump to that conclusion. But occasionally someone provides a valid interpretation that the ISS, astronaut, and M&Ms are all *falling around earth* at the same rate.

Another proper explanation would be the force of gravity is balanced by the centripetal force of rotation. Or, it can be stated that the ISS, astronaut, and everything inside follow a geodesic curve in space. Sometimes upon hearing this, lay persons and technical specialists who had answered incorrectly will say, "Oh, right. I knew that!"

Every life scenario confronts our senses, experiences, and perceptions, challenging us to make correct judgments. Why is that important? After a career developing and teaching technical training for gas turbine and rocket engines, use of the floating-M&Ms lesson demonstrates the need of effective root-cause analysis.

That process is particularly important in high-risk businesses like gas-turbine, rocket-engine accident investigation. Initial observations and reports can be misleading or taken out of context. For instance, at the beginning of a team inquiry into multiple high profile, bearing failures, a business leader declared that the affected gas turbines were assembled near an area undergoing reconstruction.

He proposed that the dust from demolition of concrete structures must have created bearing contamination. That statement generated considerable traction with some of the team, but after a few quick tests, it was ruled out and had no connection to root-cause failure. Correlation is not causation, meaning events happening simultaneously doesn't mean they are related.

Even in technical situations, most people jump to seemingly apparent conclusions without properly evaluating all the evidence. The force of gravity in the ISS stands at about 92% of earth's surface gravity, a function of altitude above earth. Because astronauts and M&Ms appear to float in air, it is easy to make a hasty assessment and say there is no gravity.

A careful observer, even without knowing the principles of orbital dynamics, should notice the ISS orbiting earth through the window. Consider the picture in the context of how we experience this world. Without a window in the ISS, there would be no simple way to determine the station was in orbit. Only a sensitive Coriolis meter could register the station going in a circle every ninety minutes.

The same principle governs the way we interpret events in our world. To properly understand evidence, information must enter from outside the immediate, observable environment, just as the ISS windows reveal to the astronaut that he is rotating around the world, despite not sensing the force of gravity.

Likewise, understanding our position in the universe requires knowledge from outside our immediate world-view. That can only occur if indeed God provides the means to communicate through that transcendent window. Second, we must take advantage of what He relates rather than closing a "blind" to the window. That window of revelation takes the form of God personally revealing Himself to anyone who believes and seeks to experience the knowledge of God.

His interaction with His creation does not bring contradiction to the general revelation of His Word. Knowing God through the Bible by His Spirit provides understanding of the world beyond the natural. While not infallible, the spiritual relationship between true believers and their Maker provides a constant check of discernment to help correct errors in life. The history of the church demonstrates the power of God's Holy Spirit to provide direction and correction through personal experience, the Bible, and other believers (His Church).

From my personal encounter, reading the Bible cover to cover did not result in God revealing Himself to me. Instead, it required personal repentance and asking God to reveal Himself before the knowledge of God became real. Suddenly experiencing the Word of God in the person of Jesus Christ brought the Bible to life.

A common objection to the ISS parallel would be that, for all practical purposes if someone cannot sense gravity, then it can be ignored. That might seem to make sense if perception equaled reality. An atheistic worldview takes such a position, even though everyone must eventually exit this life and enter the presence of God Himself. Establishing a relationship with God prevents eternal separation, subject to satisfying His conditions of salvation. Similarly for business safety, risk of injury and death are reduced when safe procedures are followed.

Outside of relationship with God, at the end of life all that remains is a collection of objects and events. For believers and skeptics, life pivots on three foundational principles of the human condition: Being, Knowing, and Doing.

1. *Being*: our identity rests on existence in relation to our Creator, the source of life and all things.
2. That must foster *knowing* truth, which involves what you know and how you know it; words have meaning.

3. Knowing leads to *doing*, which is acting with the responsibility and purpose; beings have purpose.

Everything gets flipped on its head if we try to make *knowing* and *doing* the starting point of our identity.

Space and Time

Who is there so wise,
who can indeed define,
what or where is space,
and explain the pass of time?
For so they both constrain us,
to mold the shape of life,
which we view as sculpture,
shaped by unseen knife.

Space you say is distance,
or length and height and breadth,
spanned by some contrivance,
used to plumb the depth.
Then objects be the yardstick
by which we measure space,
or they serve as anchors,
set to mark a place.

Time you state is history
tis now or future days.
So what are they but memories
that befall us on their ways?
All of time is measured,
by events that we perceive,
whether hands that sweep a clock,
or morning, noon and eve.

Objects and events,
the treasures we unearth,
Spoiled from space and time,
as measures of our worth.
How they wane to naught,
nigh His boundless scale.
In His timeless realm,
temporal notions pale.

William McGreehan (1989)

Amen, Amen!

During the final year at the University of Cincinnati for a master's degree in aerospace engineering, a rarified-gas-dynamics course was required to complete the propulsion major. Spacecraft returning to the earth's atmosphere transition from the vacuum of space to outer layers, beginning with the ionosphere where gas molecules separate due to solar and cosmic radiation. Ionized or charged particles move randomly within the region, experiencing few collisions with other particles.

Every molecular interaction becomes a tiny impact that releases momentum and energy. That may appear simple—unless you're a spacecraft or meteorite entering the ionosphere at very high velocity.

The large number of interactions creates heat, a lot of heat, without anything to dissipate heat, short of releasing heated material.

If you are a Space Shuttle or other winged vehicle, the challenge is not just cooling but decelerating and control (maneuvering). Managing this high-speed flight through rarified gas does not involve the complicated equations typical of fluid flow but a more complex type of mathematical analysis.

Scheduling my rarified-gas-dynamics class required some juggling after students warned that one of the two available professors only gave A's to those for whom he was academic advisor.

With straight A's thus far, it would be risky to take his class, but graduating on time meant doing so and I knew how students tend to inflate rumors. The risk was worth not extending school an extra semester.

Grueling would be an understatement for the mathematical analysis in that class. Trying to remember all steps of convoluted equations required extensive repetition and consumed a lot of neurons in the process.

As warned, the professor's grading was severe with every minor oversight receiving a flurry of red-pencil marks. As the final exam approached, the class got more than my usual attention, to the point of working through all of the calculations again and again to assure an A. After the final test when the Blue Book exams were graded and distributed, written in red on the cover of mine book was a failing score of 45 out of 100—impossible!

Looking through the answers, one problem's response could have legitimately had a few points taken off for being incomplete, but the professor gave it a zero.

Another major-analysis problem with the correct answer also had a zero. And so it went. Immediately after class my protests in his office were met with dismissive comments like, "That derivation had the right answer, but the terms that cancelled out happened to be misplaced. I think you just wrote some equations down and had the answer memorized. Don't worry, you will get a B for the course."

Lacking access to a blunt object, it was not worth further disagreement. Sadly, what others warned about had materialized.

Those in authority often act only out of self-interest. The same influences were in place as Jesus began His ministry.

Temple Guards

"…Never man spake like this man" (John 7:46) said the temple guard. Officers charged by the chief priests and Pharisees to arrest Jesus (John 7:32) returned empty handed, much to their leadership's dismay. In this case, dereliction of duty for temple guards merely brought a scolding:

> Are ye also deceived? Have any of the rulers or of the Pharisees believed on him? But this people who know not the law are cursed (John 7:47-49).

Temple officers occupied a distinct role apart from Herod's guards or Roman officers and soldiers. These men provided security, crowd control, and peacekeeping for Jewish rulers. As such they operated under a much more informal framework of enforcement than the others.

Temple Guards and Jesus

In the final six months of Jesus's three-and-a-half-year ministry, rather than continuing to be more reserved in His responses, He began confronting Jewish leaders openly and calling out their hypocrisy. Favorable sentiment toward Jesus had risen to such a level that Jewish leaders determined it was time to formally deal with Jesus in the absence of crowds. But temple guards sent to confront Him were so moved by the words of Jesus, they could not bring themselves to execute their assigned task.

What was so compelling? The words Jesus often began with: "Verily, verily…" Whether looking at the root meaning in Greek, Hebrew, Aramaic, English, or a myriad of other languages, He was saying, "Amen, amen…," which can best be stated as "truth" or "so it is."

Amen is an exclamatory expression to affirm a declarative statement. Does anyone today use amen as a preface or a double preface that declares "Truth, truth…" or "So it is, so it is…"? No, amen is never used before a statement in contemporary communication, nor was it used as such in Scripture—apart from Jesus.

In the Scriptures Jesus never confirmed the words of another by exclaiming amen. You won't find the expression so used, though He did affirm statements from Peter, the Roman centurion, the Syrophoenician woman, and others.

But Jesus as God used amen only when affirming his own words as truth. No further qualification was needed. After someone has spoken or written something, listeners and readers will later disagree about what was communicated. How do we resolve who interpreted the message correctly?

Often the process falls to investigation of what was said or justification of what was interpreted. During fact finding, the question to the original speaker should be, "What did you mean when you said…?"

The disciples asked Jesus what he meant by the parable of the sower (Luke 8:4-18) and Jesus commended them for asking before providing the interpretation of the story. He prefaced the explanation with:

Unto you it is given to know the mysteries of the kingdom of God: but to others in parables; that seeing they might not see, and hearing they might not understand (Luke 8:10).

At a time over forty-five years ago, auto repairs were a weekly ritual for me. Once while setting engine timing, I needed an assistant to raise and lower the engine speed for the adjustment. Debby reluctantly stepped in the car to do so, while I held the timing light and waved my hand at Debby to raise or lower the speed.

But when waved down to lower the speed, Debby did the opposite. Upon peering around the hood at her with a stern look and more frantic wave down, she sped up even more. Now quite annoyed, I forcefully motioned *DOWN!* and Debby put the peddle to the floor, pegging the tachometer until the engine let out a loud BANG and a cloud of smoke.

I yelled, "WHAT WERE YOU DOING?"

She replied, "YOU MOTIONED TO PUSH THE PEDDLE *DOWN!*"

To this day, a downward hand motion between us remains our signal for, "What do you mean?" Miraculously, the engine restarted and ran just fine after our unorthodox timing adjustment—no harm, no foul.

Much misunderstanding of God's biblical communication comes from trying to interpret or argue about what was said. Asking Him, "What did you mean?" and believing God for an answer often results in all parties recognizing they were missing the point all along. Fifty years of teaching Biblical subjects (and engineering) affirms the conclusion most people are "missing the point".

An enduring mystery of the Old Testament Levitical priesthood arises in use of Urim and Thummim. These two articles are thought to have been stones in the Levitical ephod garment used by God to reveal His commands (Exodus 28:30). Urim (meaning lights) and Thummim (meaning perfections) served the priests with God's "perfect enlightenment," revealing the Lord's will whenever the priests requested a response.

Unlike today's Apple phone Siri or the Google assistant, Urim and Thummim never misunderstood the question. Today we don't fully understand the nature of biblical Urim and Thummim—other than a method of discerning divine direction in response to a question. But as Jesus said, believers now can ask of God "and it shall be given ... seek, and ye shall find ... knock and it shall be opened" (Matthew 7:7-10). Jesus also stated:

> Fear not, little flock; for it is your Father's good pleasure to give you the kingdom (Luke 12:32).

In these final days before God separates His kingdom and the world system into sheep and goats (Matthew 25:31-33), believers need biblical discernment and the guidance of the Holy Ghost, which are the true sources of perfect enlightenment.

In Matthew 24:35 Jesus affirmed that all Bible events have been and will be fulfilled, from His past creation to His future Second Coming and the new heavens and earth. On the surface, descriptions of some events may seem difficult to understand, but God does still provide "perfect enlightenment."

Job, Elihu, and Friends

Chapter Seven - Headwinds

DURING MY GRADE-SCHOOL years, the late 1950s brought a troubling series of setbacks for our family with the sudden loss of a young sibling in a pedestrian accident, a serious automobile crash, the beginning of an unexpected and prolonged layoff, and the unanticipated death of a grandparent—followed by a parent's extended hospitalization.

Sometimes people feel like God is their opponent, especially when the first reaction of bystanders and friends is to suggest reasons why these things are happening. Even well-meaning friends and family could only offer consolation rather than meaningful answers.

Through those years Dad became increasingly angry, his bitterness particularly vented at God. But God worked a plan behind the scenes not visible until much later in the journey. Like we see in the story of Job, trouble often comes as a cascade of calamities.

Not until over forty years later in 2001—while Dad resided in a retirement facility—did he open up to hearing details of my story and experience. Upon the invitation to accept Christ, he responded, "I think I can do that." It was transformational.

Like a multi-facetted gem, the book of Job gives countless lessons beyond Job's trials and tribulations. A lengthy dialog between Job and three comforters (Job 3:1-31:40) developed out of Job's sudden series of family tragedies and personal afflictions. For a book regarded as the earliest Old Testament literature, the back-and-forth debate among Job and his friends resembles a verbal sparring match between courtroom attorneys or the prose from a Shakespearean scene.

One of Job's arguments (Job 9:1-10) reflects his awe at God's creation, while providing an outline of the ways God orchestrates actions in heaven and on earth:

"Which shaketh the earth out of her place, and the pillars thereof tremble. Which commandeth the sun, and it riseth not; and sealeth up the stars. Which alone spreadeth out the heavens, and treadeth upon the waves of the sea. Which maketh Arcturus, Orion, and Pleiades, and the chambers of the south." (Job 9:6-9)

Job declared God spreads out the heavens. The Hebrew word *natah* (naw-taw') means stretching or spreading out. An early patriarch, Job understood the skies did not always exist as observed, but during creation they were stretched out like a fabric.

Adam observed the heavens on the sixth day. Four days after God spoke the firmament into existence, did Adam eye the heavens stretched much as we see them at today? According to the Genesis record, two days after the sun, moon, and stars were made, Adam saw them defining day and night.

Living during the time of Abraham, Job excelled his peers in possessions, esteem, and godliness. Scripture depicts his character as perfect (complete). Yet, as a result of Job's exemplary reputation, he came into Satan's crosshairs.

The irony of Job's trouble was how Satan made his accusations to God about Job, only after God noted Job's integrity. Satan asserted Job's integrity resulted from God's favor and protection. So Satan received permission to remove Job's possessions and strip away his family with only his wife remaining to "encourage" him.

Even so Job maintained his integrity. Then, though Job received commendation again from God, Satan accused him and was allowed to test Job further by taking his health, though preserving his life. At this point Job's three friends arrived, seemingly to grieve with and comfort him. Instead, their support turned to counsel and then counsel to accusation.

Eliphaz the Temanite, Bildad the Shuhite, and Zophar the Naamathite variously suggested Job must have done something to displease God; otherwise, why all his troubles. They reasoned the wicked suffer and the just prosper. The three counselors advised Job to acknowledge his sin, repent, and be restored to former success.

Throughout the next thirty chapters, Job affirms his innocence while his friends debate reasons for his guilt, judging him on the basis of experience, tradition, and the law—while absent evidence. To his friends, Job's suffering seemed sufficient demonstration of God's displeasure.

Finally at the close of chapter thirty-one, Job ends his defense, content to rest his case. Although Job maintained his integrity after a series of devastating personal losses, at one point he finally decided God owed him answers, asking, "What have I done? Why does God not reason with me? How can God be so unjust?"

Ministering to prisoners in jails and prisons, I've experienced a common theme that surfaced with the ex-convict, T-bird driver while hitchhiking in the Carolinas: a tendency to justify self and find fault "out there" with others or the system. Often with them—like Job— God becomes a convenient sounding board for self-justification.

Trouble, like the characteristic four winds, arises from a variety of sources. As with our family's troubles in the late 1950s, ultimately God maintains sovereign controls of all circumstance for a purpose.

Winds of Change

Up from the depths of troubled seas,
the waves of vapor rise,
ill content to stay entombed,
when drawn to sunlit skies.
Borne by buoyant forces,
this shifting, unseen wind,
spills from a heated cauldron,
like thoughts from deep within.

Quickened ever higher,
in pursuit of clouds to fuel,
the fevered mists surrender,
as tortured currents cool.
So is set to motion
the means of change and mood;
cursed when uncontended,
yet blessed in servitude.

This full expanse of earthly orb,
Prevailing winds command;
which circle in their course,
and mock themselves again.
Now gently and alluring,
A cool, refreshing breeze,
brings joy to souls afflicted,
and sets the mind at ease.

When scolding gusts harbinger news,
of seasons change to come,
dejected spirits voice complaint,
and hopelessly succumb.
Yet vision peers beyond the gloom,
and conceives a bright milieu;
rejecting fruits of circumstance,
it serves to change venue.

Climatic patterns etched through time,
preside o'er each locale,
infusing with provincial air,
the overcast morale.
And blessed the pilgrim deft of will,
who toils while others scorn.
Not gilding weathered edges,
his tools o'er time are worn.

At times the storms beneath are loosed,
where passions have their birth,
and driven by cyclonic winds,
they rise to haunt the earth.
Who is He that rules the earth,
who wind and wave obey?
At the name of Jesus,
The plan of God holds sway.

William McGreehan (1991)

Job's friends made the obvious leap of judgment assuming he got what he deserved, despite a lack of evidence against him. Though Job did nothing overtly wrong, God expressed displeasure at him for expending energy on self-justification.

Job had forgotten his own words: "Though he slay me, yet will I trust in him..." (Job 13:15). He would have done better to declare, "Regardless of all that is happening, God is just and will work all things for good." If that sounds glib, consider who enters the picture next.

Suddenly Job's story takes a turn with the introduction of Elihu, until now a silent spectator. As a younger man than the others, Elihu has been observing the back-and-forth debate like a tennis coach, recognizing flaws in both lines of reasoning:

> Then was kindled the wrath of Elihu the son of Barachel the Buzite, of the kindred of Ram: against Job was his wrath kindled because he justified himself rather than God. Also against his three friends was his wrath kindled, because they found no answer (evidence), and yet had condemned Job. (Job 32:2-5).

From there Elihu begins a long and passionate counter to Job and his friends. Commentaries on Elihu's address are mostly critical of his arguments, rebuking him for not consoling Job. What is known about Elihu? His monologue defines his role as a divine messenger contributing as an image of Jesus Christ. Consider Elihu's list of credentials:

1. Elihu in Hebrew means "His God" or "He is my God."
2. He is younger but wiser by the inspiration of God. (Job 32:8)
3. Job's three friends were amazed and fell silent. (Job 32:15)
4. He judges fairly, even despite any man's status. (Job 32:21)
5. His words are upright. (Job 33:3)
6. The Spirit of God is in Elihu and the breath of God gave him life. (Job 33:4)
7. He speaks in God's stead, though made of "clay." (Job 33:6)
8. He is interpreter, advocate, and ransom for the lost who repent. (Job 33:21-27)
9. Following Elihu's reasoning (Job 32:1-37:24), God speaks out of the whirlwind.

10. God was angry and chastised Job's three friends while completing Elihu's arguments.

After correction from Elihu and then God, Job realized his misunderstanding and repented. Being justified and at the encouragement of God (Job 42:6), Job prayed for his friends (42:10). During the rest of his long life that followed, Job was restored by the Lord and blessed with twice what he had lost.

Though his story ends with a broad, redemptive message, Job's trials paint a portrait of God's mercy even in suffering (James 5:11). Job's story illustrates how God works in mysterious ways: allowing evil, redeeming circumstances, restoring by grace.

Early in the age of biblical patriarchs, Job looked at the heavens and declared that God "alone spreadeth out the heavens, and treadeth upon the waves of the sea." (Job 9:8). That image foreshadows Christ, who stretched out the firmament at creation, and later He came to earth walking on the sea.

And Jesus *will* come again, rending and rolling back the heavens at His return, as described in Matthew 24:29,30 and Revelation 6:12-14. With current events foreshadowing His return, the coming Day of the Lord is unfolding in our time, and when it happens His followers have already been warned to not look back and misinterpret calamities: "He that is in the field, let him likewise not return back. Remember Lot's wife." (Luke 17:31,32).

Chapter Eight - Partial Credit

URING MY SOPHOMORE year in high school, after a close call with a failing grade in geography, it had been gratifying to finish the course with a B—whether a miracle or grading by the curve. That brings up two related and commonly used phrases: "God doesn't grade on a curve" and "At least I'm better than…" Luke 18:9-14 is an example of the latter with the Pharisee thanking God that he was better than a publican (tax collector).

By nature, we tend to be competitive so self-justification comes naturally. A popular comparison might be, "I can't believe anyone would spend that much money on a car!" … or house, … or clothing."

For that mindset, spending a few years working or volunteering overseas can be an eye-opener. Much of the world's population views an automobile, home, or even an extra set of clothing as extravagant. Even adjusting to the poorer communities in developed countries can be a cultural challenge. But does God grade on a curve? No. What is His standard? Jesus said it, "Except your righteousness shall exceed the righteousness of the scribes and Pharisees, ye shall in no case enter into the kingdom of heaven." (Matthew 5:20).

To us it sounds like comparison, but with that statement Jesus raised the bar to an impossible level. God's standard is not achievable by man (except for Jesus, who was fully man and fully God). That impossibility for mankind is typified with Jesus saying it's harder for the rich man to get into heaven (on his own merits) than it would be for a camel to pass through the eye of a needle (Matthew 19:24).

Jesus brings hope through His clarification: "With men this is impossible; but with God all things are possible" (Matthew 19:26). Divine judgment cannot be based on a sliding standard of value, but with every human being a sinner, judgment must be based on the action of the individual unless they come to rely on the person and completed work of Jesus.

Through Christ, any sinner—no matter what they've done—can repent, confess Jesus as his or her Savior, gain Christ's forgiveness, and thus enter God's kingdom. And that is regardless of personal merit. In Micah 7:19, it says God will cast our sins "into the depths of the sea." And He posts a no fishing sign.

When we accept Christ's gift, God forgets our sins—and in fact He obliterates them—which means Him putting them completely out of His thoughts forever.

In God's eyes, there is value in a seeming small act of kindness. Note the distinction between justification for sin and something of value in the kingdom. Only by faith in the sacrifice of the blood of Jesus Christ and confession of Him are we granted eternal life. Born-again believers create value in God's kingdom by obedience to His will; not because God needs the effort; but rather as co-laborer with Him.

Here Jesus says it well: "Whosoever shall give to drink unto one of these little ones a cup of cold water only in the name of a disciple, verily I say unto you, he shall in no wise lose his reward" (Mathew 10:42). The following poem is a tongue-in-cheek parody about self-justification.

Keeping Up with the Joneses

You think yourself successful,
    with all you did aspire.
Whatever bought the Joneses,
    you quickly did acquire.
Despite the costly burden
and swelling mound of debt
you spent in wild abandon
and never did regret.

But all your fool expenses,
    are nothing next to me.
I climbed to greater pinnacles
    of financial wizardry.
For all my wealthy neighbors,
    thought to do me shame,
by flaunting what they purchased,
    as though it were a game.

First there were the Rogers,
    they have a brand-new car.
And so I bought a foreign make,
    with TV set and bar.
The Higgens built a pleasant home;
    a beauty to behold.
I constructed twice as big,
    with bathroom fixtures gold.

Then there were the Freemans,
whose clothes were once the rage.
So I wore the best of fashion,
    though spending half my wage.
And don't forget the Chandlers,
    whose vacations are a dream.
But I go to Europe weekly,
    or cruise the Caribbean.

Next, I saw the Bowers,
    install a pool to swim.
So now I have a yacht,
    And a lake to sail it in.
As if that weren't enough,
I charged each spending spree,
with an endless line of credit,
from a plastic money tree.

And now I sit imprisoned,
    chained by debt and greed,
still proud of what I 'complished.
    In this I did succeed.
Hear now my final boast,
    of all my worldly good.
You kept up with the Joneses,
    but I beat the neighborhood.

William McGreehan (1989)

Whether changing someone's tire, preparing a meal or helping the needy, the temptation is to focus on getting it done, but the greater value involves multiplication by being a faithful example for others, whether teaching a child or a disciple.

Jesus said, "The harvest truly is plenteous, but the labourers are few; pray ye therefore the Lord of the harvest, that he will send forth labourers into his harvest" (Matthew 9:37, 38).

One day as I returned home from work, a young lady had parked on the side of the highway, cautiously inspecting her flat tire. Seeing her predicament and the risk, I pulled over and asked if she had a spare tire and knew how to change it, to which she answered "Yes" and "No."

Having taught two daughters to drive and fix a flat tire, I said, "You're going to learn." We got started by having her notify family with my name and phone number. In the next fifteen minutes she replaced the tire while I sat on the guardrail and instructed, much to the amusement of passing motorists.

God does not need help, lest He owe something. Rather, he commands and welcomes opportunities for us to participate with Him. He looks for ways to include believers in the work of His kingdom, while letting us know He ultimately completes His own work.

When Jesus Christ returns in the clouds prior to the day of the Lord's wrath, everyone who trusts in His completed work on the cross will be caught up to be with Him in the air (1 Thessalonians 4:17).

Destruction of Edom

In the 1990s, while teaching a weekly Bible Study at work, a colleague who attended the meetings asked if the United States appeared prophetically in Scripture. In order to properly address the issue, it required reference to a few statements of the prophets.

Hidden in their declarations against the nations, are descriptions of judgment against Edom. Much of Scripture, particularly the prophets, portrays events in abstract language requiring some insight. The story of Esau (Edom) and Jacob (Israel) follows that pattern.

Jacob Deceives Esau

A hostile split occurred between Issac's two sons, Jacob and Esau, over younger-brother Jacob's deviously obtaining Esau's birthright and the blessing by deceiving their father Isaac. The brothers eventually reconciled at Jacob's return from Padanaram, yet after Israel's exodus from Egypt, a longer-term rift occurred between their offspring.

By that time descendants of Esau were referred to as Edom (meaning "red' or "earth") after his red, hairy appearance (and the red-bean soup Esau received in exchange his birthright). Jacob (meaning "deceiver") also had his name changed to Israel (meaning "he will rule as prince with God") after his all-night wrestling encounter with a man (Genesis 32:24-32), who is also described as manifestation of God or an angel (Hosea 12:3, 4).

Centuries of interaction between descendants of Israel and Edom have included limited cooperation and been mostly full of conflict. By the time of Israel's captivity in Babylon, the antagonism reached a peak as Edom cut off Israel's escaping fugitives and sold many into slavery.

Over much time Edom oppressed Israel, bringing God's judgment on the decedents of the older brother. Besides Edom, the Hebrew root of Adam is also a word meaning red or earth. By contrast, the name Israel confers a divinely appointed position of rulership with God.

In Malachi 1:2-4, God addressed the apostacy and destruction of Esau when the Lord said, "I loved Jacob and I hated Esau…" In Hebrews 12:16-17, God characterized Esau as profane (common) in selling his birthright for a bowl of soup.

Afterward Esau sought his birthright back with tears—but not repentance. The name Edom or earth adequately characterizes the worldly apostasy of Esau. Scripture warns individuals and nations who turn against His people, Israel. Two major pronouncements against Edom were prophesied by the prophets Jeremiah and Obadiah in the same era.

Both spoke from Jerusalem (Jeremiah 49:7-22, Lamentations 4:21-22, Obadiah 1-21). Ezekiel (25:12-14) prophesied similarly against Edom while he ministered to the Jews exiled in Babylon. Jeremiah's and Obadiah's condemnations against Edom drew on consistent imagery.

Thy terribleness hath deceived thee, and the pride of thine heart, O thou that dwellest in the clefts of the rock, that holdest the height of the hill: though thou shouldest make thy nest as high as the eagle, I will bring thee down from thence, saith the Lord (Jeremiah 49:16).

The pride of thine heart hath deceived thee, thou that dwellest in the clefts of the rock, whose habitation is high; that saith in his heart, Who shall bring me down to the ground? Though thou exalt thyself as the eagle, and though thou set thy nest among the stars, thence will I bring thee down, saith the Lord (Obadiah 3:3-4).

The judgment of Edom foreshadows destruction for any nation, including America, if or when they set themselves against Israel. Though God Himself and Michael the Archangel protect Israel, countless nations and world alliances have become antisemitic and anti-Israel.

The current administration in America has maintained a pro-Israel stance, which stands in the gap for the nation Israel and the Jewish people. To abandon Israel is to abandon God's redemptive plan for the nation and His Jewish people. The common refrain "God bless America" should properly be expressed "America, bless God." But that can happen only in relation to support of Israel. A remnant of God's prodigal children will eventually be redeemed as we see coming in Chapter 17 following.

As the return of Christ nears, and Antichrist comes to prominence, all nations and their leaders face unceremonious removal based on a false hope. God's two-fold promise applied to Abraham and his descendants that he would be blessed as well as be a blessing to all families of the earth. So also, all who bless Abraham would be blessed, and all who curse him will be cursed (Genesis 12:2,3).

The spirit of Antichrist threatens to subvert the world against Israel and the Church. Ultimately, all nations—including America—will face judgment, first from the Antichrist and then God. According to Scripture, when Christ returns to rule this world for a thousand years during His Millennial Reign, the only government left for Christ to depose will be that of the "The Beast" (the Antichrist system mentioned in Revelation 19:19).

Elijah and Elisha Part Ways

Chapter Nine - Three Assignments

"SURVIVE THE MONTANA Wilderness" was the challenge presented by the Public Broadcasting System (PBS) network in the early 2000s, and over 5000 applicants responded. Three families had the questionable distinction of being chosen to live in a rustic Montana homestead for five months as they attempted to meet a list of requirements. At the end of the homesteading period, each participant would be judged by experts to determine whether they were ready to survive the harsh wilderness winter. Does that sound challenging and romantic?

Despite precautions to protect the health and welfare of contestants, the show demonstrated the drama and difficulty faced for those living in a hostile environment. Participants did not consist of seasoned survivalists, but drama was created from each player having their own issues and limitations.

PBS produced similar series with families attempting to live and prosper through other taxing scenarios, always judging contestants on whether they had met certain conditions to survive or succeed. The key would be to remain focused on survival while accomplishing certain tasks and avoiding unnecessary distractions.

Elijah had to learn similar skills. He stands out as the greatest Old Testament prophet, even though his apprentice Elisha received a double portion of Elijah's anointing and performed twice as many recorded miracles—Elijah's seven versus the younger Elisha's fourteen. When Jesus declared John the Baptist greater than all the prophets (Matthew 11:11), He noted that John came in the spirit of Elijah (Matthew 17:11-13).

Yet, after his miraculous challenge, confrontation, and defeat of eight hundred and fifty prophets of Baal and Asherah on Mount Carmel, Elijah fled in fear at Jezebel's threat to have him executed (1 Kings 19:2-3). Not long before this Elijah had been hiding in the wilderness and was fed by a spring and ravens bringing bread and meat (1 Kings 17:6). I wonder if Elijah questioned, "Is this kosher?"

After all God had done through him, suddenly Elijah's confidence wavered and he fled. A fugitive again, this time the Lord's angel fed him. Elijah had run a long way before arriving at Horeb, the mount of God, where he hid in a cave.

There the Word of the Lord came to Elijah by a still, small voice, "What doest thou here, Elijah?" (1 Kings 19:9). After Elijah lamented to the Lord about his perceived state as the last faithful prophet in Israel and that Queen Jezebel was threatening his life, the Lord demonstrated His power for Elijah, including calamitous wind, earthquake, and fire.

Rather than speak to the prophet from the chaos, God repeated His question in a still, small voice, "What doest thou here, Elijah?" With that, Elijah still restated his complaint, so the Lord commanded him to return the way to the wilderness of Damascus and execute a three-fold assignment (1 Kings 19:15-16): anoint Hazael king over Syria, Jehu son of Nimshi king over Israel, and Elisha son of Shaphat to take Elijah's place as God's prophet.

Given Elijah's eventual replacement, you'd think his focus would have been completing all three tasks; indeed, Elijah immediately found and anointed Elisha as his prophetic replacement.

So Elijah completed only the third task, and the relationship with the two prophets continued for roughly ten years before Elijah was caught up to heaven by a chariot and horses of fire. Then a double portion of Elijah's spirit fell upon Elisha (2 Kings 2:11), as he'd requested of Elijah when taking over his position. Elijah said he would get the double portion if Elisha saw Elijah taken away.

Undoubtedly Elijah delegated his remaining two assignments to Elisha, yet it was about fifteen years later in his long prophetic ministry that Elisha completed the second task of anointing Hazael king of Syria (2 Kings 8:7-15) and twenty-two years later that Elisha delegated an unnamed son of the prophets to anoint Jehu king over Israel (2 Kings 9:1-3).

Did Elijah and Elisha procrastinate on the last two assignments? Examining biblical events in Syria and Israel, Hazael and Jehu were not yet in position to take their respective thrones until many years after God's instructions of anointing to Elijah.

Scripture contains numerous examples of those in authority accomplishing responsibilities by sending a messenger or servant, with a classic example being the centurion's ill servant. As related in Matthew 8:5-13, a centurion from Capernaum personally made request to Jesus.

The corresponding account in Luke 7:1-10 says the centurion sent Jewish leaders to Jesus with his request, after which he then sent friends to appeal that Jesus would simply "...say in a word..." and the servant would be healed. In both portrayals, Jesus commended the centurion for his faith in Jesus to accomplish the result simply by speaking the word subject to God's authority. Not only did the centurion act under Roman authority, he also knew a word spoken by faith under divine authority equated to the act being done in person.

In the case of Elijah's three-fold assignment, the Lord completed His Word; first by commanding Elijah, then Elijah passing two tasks to his servant Elisha, and finally Elisha picking an unknown prophetic-ministry student to carry out God's last command. The three each completed and—or delegated one of God's three assignments.

Jesus expanded the principle of authority by faith, declaring John the Baptist was greater than all the prophets—and yet, under the new covenant, the least believer holds a greater authority, "...he that is least in the kingdom of heaven is greater than he [John the Baptist]"

(Matthew 11:11). Jesus said that we as Christians would also do great works:

> Verily, verily, I say unto you, He that believeth on me, the works that I do shall he do also; and greater works than these shall he do; because I go unto my Father. (John 14:12)

Today, every follower of Christ wields greater influence than John the Baptist or Elijah by speaking the word faithfully and in authority. I often remind others, "Everything is easy to the one who has done it", or the corollary, "Everything is easy to the one who doesn't have to do it".

This final, brief season will pass quickly into what the Bible calls the Day of the Lord, when neither great nor least will toil or work. The Prophet Isaiah declared about that day:

> And it shall come to pass in the last days, that the mountain of the Lord's house shall be established in the top of the mountains, and shall be exalted above the hills; and all nations shall flow unto it (Isaiah 2:2).

In Isaiah's verses below, and then through to Isaiah 5:30, the same expression stated "Day of the Lord" or "in that day" appears eight times, along with the five woes of chapter 5:

> The lofty looks of man shall be humbled, and the haughtiness of men shall be bowed down, and the Lord alone shall be exalted in that day. For the day of the Lord of hosts shall be upon every one that is proud and lofty, and upon every one that is lifted up; and he shall be brought low... (Isaiah 2:11,12).

References in Scripture to the Day of the Lord are about His wrath to come and no mention of any works of man or Satan. At that final stage of history, God alone will be centerstage, and He promises to close the curtains and dim the lights.

Mary and Joseph Find Jesus with the Teachers

Chapter Ten - The Script

EW THINGS IN life equal the torment of lying in bed, staring at a night-time ceiling, after having realized you need to make something right. While regularly ministering at Butler County Jail, that hook seemed to catch everyone's attention when I framed it like this, "Did you ever get away with something that no one knew about, but then you lay awake in bed at night, unable to sleep until you tell someone what you've done?"

Immediately the room would let out a collective gasp, "Oooh, yes!" Many a jail-house snitch has fished at that well to catch leniency. Within the soul of everyone, God created the tension of knowing when something needs to be addressed. Whether in a public courtroom or the trial of conscience, the choice comes down to awaiting judgment or pleading for mercy. In both cases, it helps to have an advocate on your side.

During his second forty days on the mountain (Exodus 32-34) and after Aaron and the people idolized a golden calf, God warned Moses the people had corrupted themselves. The Lord let Moses know how He planned to judge them:

> Now therefore let me alone, that my wrath may wax hot against them, and that I may consume them: and I will make of thee a great nation (Exodus 32:10).

How should Moses have responded? It seemed God had already made up His mind, and on the surface, this sounded like a promotion for Moses. Instead, Moses stepped in to calm God down (verse 11) and remind Him who He is (verse 12), what He promised, and how the Egyptians might interpret this.

God relented at the intervention of Moses and Bible skeptics have a field day with a description like this. It appears Moses' calm reasoning prevailed over God's anger. What would God have done were it not for Moses? Those who know God personally recognize He creates apparent riddles, but always acts consistent with who He is, and that means sin must be judged.

God doesn't threaten, "Don't make me count to three" or excuse that He "didn't really mean it." Instead, to resolve the tension He declares judgement, while looking for an advocate to plead for mercy.

In the courtroom, sentencing after a conviction generally follows prescribed guidelines to avoid arbitrary punishment. Either sentences are based on jury recommendation, mandatory minimum, or prescribed by circumstances of the crime.

Though a judge does not always have discretion, he can make exceptions based on an attorney's plea. At sentencing, and acting as his client's advocate, a defense attorney can petition the judge with a plea, "Your honor, there are some extenuating circumstances concerning my client's situation."

The judge can then honor the advocate by reducing a prescribed punishment and still be satisfying his judicial responsibility. That is mercy. For those who need an advocate in life, Jesus stands ready like Moses (just as a judge will always appoint an attorney if needed).

While it's not proper for a judge to say, "We'll overlook this," submitting to an advocate's appeal for mercy is. Our form of judicial rule follows God's design—at least it was intended to—and when He passed before Moses God declared:

> The Lord, The Lord God, merciful and gracious, longsuf-fering, and abundant in goodness and truth, keeping mercy for thousands, forgiving iniquity and transgression and sin, and that will by no means clear the guilty... (Exodus 34:6-7)

God shows mercy but does not clear the guilty as though it never happened. There are consequences for sin. Someone must pay the price. And indeed, Christ did:

A Perfect Ten

Pick a number one to ten,
Rate yourself with God and men.
Do all others call you fine?
You are worth at least a nine
Were you born of noble fate?
Give yourself no less than eight.
Are you mindful oft of heaven?
God might rate you close to seven.
Those who problems help to fix,
    Are rewarded with a six.
One who lives but to survive,
    can expect to earn a five.
Do a job and nothing more,
    get respect and just a four.

Complain to friends of all you see,
    and they will give you only three.
Rush to court intent to sue,
    win or lose achieves a two.
If of good your life has none,
    you are left with but a one.
Last of all the losers' hero;
    nothing but a lowly zero.
There's a ONE who paid the cost.
HE came to seek and save the lost.
    Become a zero so that when,
    you're with HIM a perfect ten.

William McGreehan (1991)

As He taught Jonah with the withered gourd (Jonah 4:1-11), God demonstrates longsuffering and mercy by prolonging judgment. The inhabitants of Nineveh, capital of Assyria, repented after Jonah's preaching, and much to his dismay, God showed them compassion. Yet one hundred fifty years later at the prophesying of Nahum (1:1-3:19), the Lord's wrath fell on Nineveh and the rest of Assyria. Likewise, after Jesus returns in the heavens to catch away believers, the Day of the Lord's Wrath will begin to execute final judgment on earth. Until then, mercy prevails (James 2:13).

After years teaching adult ministry and training engineers, along came grandchildren. With two grandsons, Andrew and Noah, in a faith-based scouting program, opportunity arrived to help out during their scouting events.

With children and grandchildren, it bears remembering that we're raising adults—not children. Roughly half the boys in the program lacked a father at home, and even fewer had access to someone who could teach basic life skills.

Every project and activity required a balance between what we did as leaders and what would be delegated to the boys. For derby cars, we leaders planned and roughed out our scouts' pinewood blocks and then they were responsible for finish sanding the car bodies, painting them, and installing the wheels, all the while able to ask for our assistance when needed (ownership comes with investment, not entitlement).

Derby Car Project

Once during a camp outing, a boy who would have rather just played in the woods asked, "Why are you having us do all the work?" I replied, "We already learned how to do this. When you become a leader, make your boys learn it."

Another skill development involved creating a collectable knife. We arranged for the boys to make the handle and then connect it to a steel blade. After hearing the personal testimony of Brad Vice, founder of Alabama Damascus Steel, we decided to purchase his Damascus Steel knife blades for the boys. But while discussing the plan with Brad, he donated blades, and the boys learned how to assemble and finish a fine knife.

Damascus Steel Knife Project

Jesus with the Teachers

Near the age of boys in our scouting program, twelve-year-old Jesus traveled from Nazareth to Jerusalem with his parents to celebrate Passover (Luke 2:41-50). On the family's journey back to Nazareth,

Joseph and Mary panicked when they realized Jesus wasn't with the extended family.

They turned back to Jerusalem and after three days of searching they found Him in the temple among doctoral teachers of the law, listening and asking questions. Scripture records that all who heard Him were astonished with His understanding and answers. Jesus posed questions, and when scholars of the Jewish law could not give satisfactory responses, He answered His own questions, much to their amazement.

The Queen of Sheba ("oath") came to Solomon, drawn by his reputation for wisdom (1 Kings 10). Depending on the translation, Solomon either told her all her answers or even the questions. The Hebrew says he told her all her issues (*dabar*), a word that means to speak or arrange. As with many Hebrew words, *dabar* has a variety of meanings, depending on context.

In news interviews and congressional hearings, the interviewers often demand, "What is your answer, yes or no?" This response is what the Jewish leaders expected from Jesus, regarding whether the law permitted payment of tribute to Roman Emperor Caesar (Matthew 22:21).

It appeared they had Jesus on the horns of a dilemma, facing criticism with either answer. Jesus instead called for a tribute coin (denarius) with Caesar's image and inscription. He declared they should pay Caesar what is Caesar's and God what is God's. Effectively Jesus said, "You're missing the point." Many challengers to Jesus sidestepped the real issue because of their misplaced priorities.

Study of Scripture and our relationship through the Holy Spirit enlightens believers' hearts and minds to understand the real issues between God and man and between man and man. If you listen to the media, science, politics, or any so-called experts, opinions on topics of concern usually miss the point.

Everything begins with us taking personal responsibility for—or failing at—our relationship with God (Matthew 22:37) and others (Matthew 22:39). The real problems are not "out there" (someone else's fault). They're "in here" (in our hearts); though, solutions do not come from within us (our thoughts and knowledge), but from revelation found "out there" with the Word of God and the Holy Spirit.

Man's earthly experience and knowledge cannot govern the interpretation of God's Word. Instead, God's Word and the Holy Spirit inform us of how the human experience must be understood. Looking back to the illustration of the window in the ISS, just as information must come in through the window to interpret the situation, God provides a means communicate the real issues and solutions.

Likewise, there is no experience to draw upon for Scripture's unprecedented, future events, like soon coming day when believers will be caught away before the Day of the Lord. Jesus declared:

> Ask, and it shall be given you; seek, and ye shall find; knock, and it shall be opened unto you: For every one that asketh receiveth; and he that seeketh findeth; and to him that knocketh it shall be opened (Matthew 7:7,8).

Eventually, secular cosmologists and even theologians will understand that the riddles of an expanding universe and the formation of life itself was revealed from the beginning. At Christ's second appearing, their oversight of misinterpreting the script will be obvious.

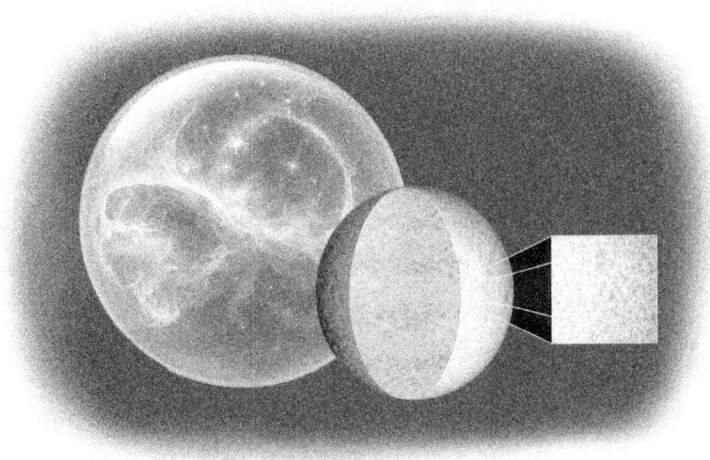

A Bounded Universe and CMB Radiation

Chapter Eleven - Opening Scene

"IN THE BEGINNING God created the heaven and the earth." Genesis 1:1 tells us the universe had a beginning, and God as the cause behind its' origin. Though not expressed in today's scientific terms, this simple, biblical explanation of the origin of the universe and life enjoyed broad acceptance throughout most of recorded history. The following background lays groundwork for the coming events at the end of the age and Christ's return.

In recent decades, a growing body of scientific knowledge has been developed to explain celestial observations using natural theories. As naturalistic explanations gained acceptance and credibility, the general public and scientific community have increasingly embraced science and the idea it has displaced belief in the creation story.

A feature of observational science includes the need to refine and at times replace theories. With time and discovery, flaws often appear in models, bringing the scientific community to a crisis requiring new insight to revise or replace a concept.

Historical paradigm shifts in cosmology have occurred regarding shape of earth, orbital mechanics, star formation, characteristics of electromagnetic radiation, the nature of spacetime, and more recently, evolution of the universe. Change brings controversy. At times transformational ideas gain acceptance slowly; other times abruptly.

For most of recent history steady state (non-expanding) cosmology maintained broad acceptance in the early twentieth century until a 1929 publication of Edwin Hubble's (1889-1953) data inferring expansion of the universe based on observed redshift as a function of increasing distance. During the last half of the twentieth century, "big bang" cosmology gained further support with telescope and satellite observation based on refinement of measured redshift with distance.

Relying on redshift, cosmologists attempt to fit data into a big bang model, where the universe began with the rapid growth of matter, energy, and spacetime expanding into a void. However, a new crisis has arisen resulting from calculated mass of observed matter in the universe that cannot be reconciled with redshift inferred acceleration.

To resolve the disconnect between observed redshift and lack of expected mass, dark matter and dark energy have been theorized. Astronomers estimate normal, observed matter to be 15% of total required mass so the proposed dark matter would make up the other 85%.

Depending on standard of measure used to estimate distance, current calculated expansion rate of the universe is determined to between 67.4 and 74 kilometers/sec/megaparsec, a range of values often termed Hubble tension. Dimensional units "km/sec/megaparsec" describes increase of expansion rate with distance. Estimated total observable expanse of the universe is 28 billion parsecs diameter (93 billion light years), or 14 billion parsecs radius.

In the past few decades, modern cosmology has come to a paradigm crisis. That is particularly apparent with results from the James Webb Space Telescope (JWST) showing the universe does not match the standard model of cosmology. How does this result compare with what the Word of God says in the creation account.

The important consideration is: What does the Word of God say about means of formation of the cosmos?

In the beginning was the Word, and the Word was with God,

and the Word was God. The same was in the beginning with God. All things were made by him; and without him was not any thing made that was made (John 1:1-3).

Jesus Christ, the Son and Word of God, spoke all things into existence, which includes the entire universe with its worlds, spacetime, and all living things. Mankind had their beginning by the Word of God. God said, let there be; and there was.

Divine decree comes of the spoken word *hayah* ("to exist" or "to be"), which is calling something into being or declaring its existence. God declared himself to Moses in Exodus 3:14: "I AM THAT I AM," or "*hayah asher hayah*" ("I exist because I exist") that speaks of underived (originated) existence. So then, all existence flows out of God's person in His divine Hebrew name YHWH (four letters called the "tetragrammaton" in Hebrew), which is the origin of God's name "Yahweh." As the writer of Hebrews declares:

Through faith we understand that the worlds were framed by the word of God, so that things which are seen were not made of things which do appear (Hebrews 11:3).

More than claiming some miracle, that verse lays a framework for the cause of all things. Does Scripture communicate the mechanism behind the heavens expanding as observed today? The Bible contains at least ten references to the heavens being stretched out by the Hebrew *natah* (meaning stretch out or bow down).

Variants of the Hebrew word *natah* appear at least 214 times in the Old Testament. On the second day, the Genesis 1:6-8 account describes creation of a firmament or Hebrew *raqia* (meaning expanse) in the midst of the waters or Hebrew *mayim* (meaning waters or fluid). God called the firmament heavens or *shamayim* (meaning lofty) between the waters speaking of the universe bounded above by an expanse of waters. The origin of the Hebrew word *raqia* comes from *raqa* (meaning to spread out or stretch), reflecting the stretching out of the heavens by the waters at the outer perimeter.

This picture of matter at the outer bounds of the heavens provides relativistic (pertaining to Einstein's Theory of Relativity) means for stretching the fabric of spacetime during God's creation, as well as explaining the redshift we see. More importantly, this same mechanism explains how a future sequence of scriptural events will happen when the heavens rend open at the return of Jesus Christ.

During the rending of the heavens, the sky rolls back as a fabric, precipitating the appearance of cosmic signs before Jesus Christ returns in the heavens to catch away believers before the outpouring of God's wrath on the Day of the Lord that follows.

However, this scenario departs from our world's two-dimensional, fabric analogy because spacetime has a four-dimensional form (three space and one time) dictated by the influence of localized mass like earth, as well as balance of matter in the universe. Universal mass including local, distant bodies, and the boundary of the heavens contributes to the concentration and local perception of spacetime. Removal of the influence of mass bounding space diminishes the impact during the events of the sixth seal in Revelation 6:12-14.

As the heavens rend and roll back, signs of Matthew 24:29-31 and Revelation 6:12-14 unfold on earth and in the heavens. The concept should not be mistaken as deterministic—in that it depends on a necessary naturalistic cause and effect outcome—rather it is a consistent, connected, concise summary of the prophesied events resulting from the rending of the heavens in Psalm 18:9,10, Psalm 144:5, and Isaiah 64:1, fulfilling the promised catching away of believers at the appearance of the Lord Jesus. More about these events later in Chapter 17 of this book during the discussion of the cosmic signs.

Population Statistics

After the 1986 Challenger Space Shuttle disaster, a team of scientists contributed to the Rogers Commission Report that summarized causes leading to the tragic failure. Dr. Richard Feynman's section of the report—regarded as a political hot potato—criticized NASA leadership's handling of safety management and was relegated to an appendix.

During the investigation press conference, a simple demonstration using an O-ring and a glass of ice water explained more than all the report's cause-and-effect, fault diagrams. Most missed the obvious that Dr. Feynman, a Nobel Laureate in physics, illustrated by placing an O-ring in ice water and then showing how it would break when flexed.

The O-ring in question was made of the same elastomeric material used to seal sections of the solid-rocket-motor casing. Exposed to overnight freezing before the fateful launch, Challenger's O-ring failed, resulting in a catastrophic leak of hot gas, impacting the liquid hydrogen/liquid oxygen tank which then exploded, destroying the shuttle and its occupants.

When it comes to question of origins, it is possible to examine evidence of human history and the present while missing the obvious. One of the primary means needed to support the theory of evolution would be very long periods of time.

Clear evidence of recent human origins (less than 10,000 years) would negate evolutionary theory as taught today. Also, evidence of evolution must be shown through changes in human genetics, such as regular mutations, some of which must be favorable.

Mutations that deteriorate the genome, only losing information, would be consistent with our world deteriorating after the fall into sin. But favorable mutations (additions of information with added and beneficial genetic change) have never been scientifically proven or observed in nature, despite suggestions of plausible improvements.

Supporters of Darwinian evolution need a universe billions of years old with hundreds of millions for living things, millions of primates, and hundreds of thousands of humans. Current reasoning and most homo-sapien studies put people as having developed within the past 50,000 to 100,000 years, which has interesting implications:

Since large populations are required over these long periods, prehistoric inhabitants would need numbers much greater than even hundreds of millions. Though human population genetics demonstrate considerable diversity, no fundamental change in the human genome akin to speciation can be evidenced within recorded history since roughly 5000 BCE.

There has been variation, but even Neandertal, Cro-Magnon and other classifications of human remains do not annul a lack of necessary observable change. This absence of continuous change in the fossil record contributed to evolutionary concept proposals such as "punctuated equilibria" by Dr. Stephen Jay Gould (1941-2002). Also termed "hopeful monsters", the concept sought to explain the fossil record with stasis or lack of change, and then sudden transition to a new species with little or no intermediary fossils.

Such evidence is, however, clearly consistent with the biblical record of creation with each species reproducing after its kind. Consider the Bible's record of human population. For the sake of discussion, there is no need to estimate population numbers prior to the global flood documented as happening 2348 BC. Post flood, the world population restarted with eight people.

Consider non-biblical, naturalistic assumptions about population dynamics. One requirement for evolutionists' genetic mutation, selection, and resulting variation would be a reasonably sized population. And they consistently argue for that speciation change having to happen through large populations over hundreds of thousands or millions of years.

But what do we observe through recorded history? A simple experiment on human populations can be used to frame the problem, which might seem difficult from estimated or inferred births and deaths from population statistics.

Instead, we'll use the same data assumptions available to all evolutionists, the latest Population Reference Bureau (PRB) report (2022). Originally produced in 1995, the PRB provides a summary of human population, presuming speciation from 50,000 BCE (Before Christian Era) to the present and estimates to 2050 CE.

Although reasonable population estimates and statistics from the past few hundred years are documented, data for recent decades are considerably more accurate. Anthropologists routinely use these population statistics in support of the evolutionary worldview.

Results are summarized in the report, "How Many People Have Ever Lived on Earth?" by Toshiko Kaneda and Carl Haub (PRB 2022). The projected total of people living before 8000 BCE (Before Christian Era) is 1.14 billion. From that, 117 billion is the estimated total who ever lived as of 2022.

These numbers create a fundamental problem. Apart from superficial variation, there has been no major genetic or observable transformation of homo sapiens since 8000 BCE. But according to the evolutionary theory of human development, basic speciation must have occurred prior to this time, which would be only 1% of total historical human population estimates to date.

Evolutionary apologists might argue that environmental and social factors stressed pre-8000 BCE (Before Christian Era) populations to more quickly select genetic mutations. But any student of human history can see that every era has had significant pressures from war, famine, disease, and environment.

Population numbers, human history, and lack of fundamental difference in human makeup for recorded history do not support evolutionary models for origins. Variations occur within genetic makeup, but there has been no positive, fundamental addition to human genetic information.

Even a cursory review of Greek sculpture, ancient metallurgy, early literature, astronomical observation, mathematics, and invention attests to the advent of recorded history showing fully formed humans with sophisticated rationality. Evidence from human genetic history does not support the change needed to speculate something different from divine creation.

Popular and scientific articles about origins frequently frame conclusions using "could have," "may have," "likely," or "the accepted explanation." Though a line of reasoning may sound reasonable, under rigorous scrutiny it doesn't hold up. My classic response in technical root cause investigation is, "Just because it's plausible does not make it causable".

Creating a universe by undirected beginning and then formation of life through time, chance, and natural selection may sound plausible with verbal gymnastics, but these cannot be shown to be an efficient cause; that is, all those influences causing specified complexity. Again, Occam's Razor demands the simplest explanation is usually the correct one, and cuts through sophisticated plausibility arguments.

Naturalistic teachings on cosmic and life origins stand "...as a bowing wall ... as a tottering fence (Psalm 62:3)," or "...the wall that ye have daubed with untempered morter..." (Ezekiel 13:14).

As the Lord declared, "The wall is no more, neither they that daubed it... (Ezekiel 13:15)." Before His sudden return to earth, a humiliating end awaits naturalistic doctrine, just like the collapse of a leaning wall. Nations, symbols, dogmas, and their proponents are all destined to fall before the presence of Christ, as the image of false-god Dagon bowed before the ark of God (1 Samuel 5:3). Error can run a swift sprint, but truth dominates the finish line.

The road trips we've taken throughout this book show life's troubles arise from many different directions. The root cause of the problem does not originate with business, labor, academia, media, churches, government, family, or neighbors.

Since the Garden of Eden, humans have been no match for the common adversary Satan. He himself is no competition for God, Who sketched out His plan from the beginning and always has been executing it perfectly and faithfully toward completion—despite resistance.

Paraphrasing how God might characterize the situation, I'll say it's much like my advice to a grandson when we made a pinewood-derby car for our church-scouting troop: "Just work with me. I've got this."

A Check for a Million Dollars

Chapter Twelve – Multiplication

DURING A NEIGHBORHOOD get-together some years ago, an acquaintance was discussing his local business venture. As he related the story, some years prior his boss and business owner in another city called him to his office.

After exchanging pleasantries, the owner slid a check across the desk for one million dollars. My neighbor said to the boss, "What is this for?" Notice the recipient's first comment was not, "Thank you, now I can pay off the mortgage, "This means I can buy a boat!" or "I can finally take a family vacation." Rather it was to ask the purpose for the money. The owner replied, "Go to Cincinnati and open a branch of the business."

After Scripture's Mount Olivet discourse regarding His second coming, Jesus told several stories portraying the kingdom of heaven. Matthew 25:14-30 depicts servants, their master, and his goods. Jesus relates how the master took a journey to a far country, but before leaving, he handed each of his three servants some of the master's money in "talents" (a very large sum).

Bear in mind, like many others, this parable (or puzzle) was not intended to explain the meaning, but to create a mystery. Once when Jesus's disciples asked the purpose of His parables (Matthew 13:10-17), Jesus replied with the example of Isaiah 6:9,10 that says most eyes of listeners are spiritually blind and deaf to their meaning. Jesus added, blessed are those like the disciples who would ask and have the meaning revealed to them.

So in the story of the talents, you might ask what man would leave town and give his servants a large sum of money. In this case it was someone who understood the meaning and value of wealth. Even knowing your servants' character and abilities, there are significant risks involved. But by investing, many of us do the same every day.

Consider the following characteristics of the talents. In going to a far country for a higher purpose, it can be implied the master could not take his wealth with him. The choice was to hide it, bank it, or invest with someone. Hiding wealth makes it worthless as we will see. Given he knew the track record of his servants, he invested in them according to their abilities.

Observe a few aspects of the investments. The man did not tell his servants what to do with the talents, yet there was an implicit under-standing, because two of them immediately took their talents and began to buy and sell with the intent of achieving a net gain.

Upon his return all three servants were called to give an accounting of what they had done with his funds he entrusted to them. A similar parable by Jesus (Luke 19:11-27) frames the story of a nobleman who departed to a far country to receive a kingdom. In both parables the main character returns with little concern for the original investment as the servants are not asked to relinquish their talents (or pounds), nor their gains.

One at a time, servants came to state what they received and gained. As a result, the master congratulated each and rewarded them according to their gains with significantly greater authority. Gains were overshadowed by the rewards and the opportunity to govern a kingdom under the master.

The only exception was the servant who did not invest. Rather he claimed to be afraid as the master was harsh. His approach was to hide the talent in hopes of preserving it. At least, he reasoned, it could be returned without risk of loss—no harm, no foul.

The last servant must have been surprised by the master's response. Luke 19 says the master told his servant that he would be judged by his own mouth. In other words, if the servant was afraid his master—"reaps where he has not sown"—the servant should have at least invested the talents in a bank to gain interest. So the master did not validate the servant's thinking or even acknowledge his fear; rather, in Matthew 25 he characterized the servant as wicked and lazy.

There are four basic actions that can be done with a medium of exchange (money): *invest, consume, preserve,* or *waste.* The first two create value, and the others destroy it. Consider the meaning of each concept.

Invest generally infers something provided as a value exchange for the purpose of asset appreciation or resale. Remember, money has zero intrinsic value; only when used can value be created. In that respect purchasing can be a form of investment. Like someone who buys a luxury boat at a bargain, lists it for resale at a significant gain, and uses it for recreation until the right buyer comes along.

Or a person could buy twenty-four water bottles, take them to an event, and sell them for five times the investment (hopefully with a vendor's license). These examples are value or wealth creation by investment, which involves buying and selling—as well as risk.

The servants in the story who bought and sold did so with confidence they would earn an increase, despite risk. A large number of seeds sown will always produce an increase. An individual seed may fail to reproduce, but with many sown liberally, there must be increase because seeds naturally grow.

Offering something of value as a donation is also an investment in an important sense. The master in the story gave each servant talents without expectation that the money or the gain would be returned back to the master. But he called them to account of their stewardship and rewarded them accordingly. Only one disappointed him when he tried to preserve his talent, because he was wicked and lazy.

A donation comes with stewardship. If a recipient misappropriates a gift, then at least we stop giving to them. A responsible investor (or giver) would not continue to invest, knowing the recipient is using it for something other than the purpose for which it was intended.

Also, purchases for *consumption* or use have value in that it is desirable or necessary for living. If there were no consumption, every transaction would be part of a grand pyramid scheme with each step inflating value, though producing no goods or services. A commodity purchased for consumption creates demand which fuels economy. Food, clothing, housing, transportation, entertainment, etc. are all valid uses for consuming resources, though they should not become an end in themselves—that's waste.

Preserving is not to be confused with saving, which involves investment to gain interest. The wicked and lazy servant tried to preserve the single talent which was intended for investment. Instead, he feared taking a risk and losing it, assuming he could preserve and return it with no fault. Money preserved under a mattress reflects laziness, not thrift. The master demanded the talent should have at least been given to the bank (or money changer table) to gain an increase.

And *waste* implies taking something like the talent capable of increase and expending it for nothing of value. For example, in the story of Jacob and Esau, Esau sold his birthright to Jacob for a bowl of bean soup and so was rejected by God. When Abraham requested a burial field for Sarah from the sons of Heth (Genesis 20:7-17), Ephron offered his field worth four hundred shekels of silver, at no cost to Abraham.

Instead, Abraham measured out the value in silver and gave it to Ephron, refusing to accept something for nothing, thus negating its value. Similarly, when Araunah the Jebusite offered King David a threshing floor, oxen, and wood (2 Samuel 24:24) for David's sacrifice, the king refused to take something for sacrifice to God at no cost himself.

In the parable of the talents, the last servant didn't understand that the point of the exercise was not to enrich the master or preserve his money: it was to enable the talent to do what it would do naturally if circulated through value exchange. This servant entirely missed the purpose and then slandered his master's motive—wicked and lazy indeed!

Seeds sown require no further action to produce results. Certainly, the yield or gain increases with watering, weeding, and fertilizing. But in the plan of God, a seed sown will sprout, grow, and produce increase because God created it to reproduce after its kind. Sowing seeds and investing in God's kingdom also increases. The plan for Christians regarding the kingdom of God involves several key actions that can be represented by the vowels: A, E, I, O, U, Y:

1. A, invade: Jesus called disciples to go make disciples throughout the world.
2. E, invest: Take what you have and invest something in the kingdom.
3. I, invite: Ask the lost to receive the free gift of eternal life.
4. O, invoke: At the least, pray for all and for the world.
5. U, you: The calling goes to all believers, not just a few, and that includes you.
6. Y, why: Jesus promised it, so the Father's pleasure is to give us the kingdom.

All believers are called to *invade* the world, whether in person, in proxy, or in print. That process fulfills God's command for us to be the salt and light, salting the world with the person of Jesus and His gospel message.

Investing has strategic value for influencing individuals and the church, culture, and world. Multiplying yourself takes place when you invest biblical wisdom in all kinds of people, yielding dividends for the kingdom. The church as the Body of Christ teaches the world while always *inviting* non-believers to faith.

Fourth, all believers can *invoke (pray to)* God about the needs of others, even if the Christian is confined at home or praying for someone on the other side of the world. Without prayer, a Christian's walk and relationship with God would be hollow.

As far as *you,* God needs no one but everyone needs Him to truly live and fulfill their life's purpose, co-laboring to spread the gospel.

The *why* should be self-evident to every believer, understanding that the Lord made you for relationship with Him. The King of kings is going to lavish His heavenly kingdom on you and who trust in Him.

Like the master returning in the story of the talents, Christ will come again, first in the air above earth where He'll catch all believers away to be with Him. Following that, all God's saints will return to rule and reign a thousand years with Jesus.

And every Christian servant of His will be rewarded authority to rule, the scope of which will depend on the trust to enable the Master's talents to do what they accomplish naturally, to reproduce. Believers are called to use whatever resource they have been given by God—whether little or much— to earn fruit for His kingdom. Invest your time and resources wisely, with the knowledge that Christ's gospel message is "always safe and earns dividends".

Principle of Multiplication

Tithing as a principle draws strong opinions, particularly from those who misinterpret the concept. What is the origin of tithe? The Hebrew-word *awsar* means tithe and derives from the word for "accumulate" (or "ten"), and those reference an "order of magnitude" or "multiplication." As part of the tithe concept and for those giving, God promises blessing and multiplication for those investing in God's purpose.

When Abraham returned with the spoils of battle, he acknowledged Melchizedek, King of Salem (peace), priest of the Most High God with a tithe (Genesis 14:18-20), who was also a picture of Christ. The literal Hebrew meaning of Melchizedek is King of Righteousness as explained in Hebrews 7 showing a portrayal of Christ.

So also, when Jacob dreamed at Bethel, he promised to honor God's blessing by giving back a tenth (Genesis 28:20-22) of his increase as tithe. God blesses what is invested in His kingdom. Like Abraham, Isaac, and Jacob who by faith sowed back into God's purpose; the sower becomes co-laborer with God in His work. You name the investment and God blesses and multiplies it, putting zeros on the end of the amount—adding orders of magnitude to it.

After retiring from the aviation gas turbine industry and joining an engineering consulting firm in Florida, a contracted position became available for a classified, hypersonic-engine project. After negotiating my role, Debby and I made a temporary move back to Florida, forty-two years after having left ministry in Florida and Haiti.

Once on the job and details of the secret hypersonic-propulsion project were disclosed, the scope involved a high-profile effort to complete engine-system integration and altitude, engine test at hypersonic speed—a dream job of propulsion engineering. Upon completing design, test planning, and preparation, the hypersonic, engine-demonstrator version was successfully operated in a high-altitude hypersonic test cell.

After completion of the testing, existence of the project was publicly released. A recent movie *Top Gun: Maverick*, with Tom Cruise as aging Navy pilot Pete "Maverick" Mitchell, opened featuring Hollywood's computer-generated version of the hypersonic dual mode scramjet powered aircraft called Darkstar. Though the engine speed was classified as greater than Mach 5, Hollywood film creators inflated the concept to a Mach 10 fighter—to sell tickets.

Hypersonic Aircraft Development

Following the hypersonic project and over the next seven years, contracts became available on various liquid- and solid-rocket-engine programs for several manufacturers. Some of these roles required creation and publication of large documents involving hundreds or thousands of pages for engine test and qualification.

In most cases, a team of technical specialists and contributors were tasked with writing and providing their sections for final documentation. Often completion necessitated rewriting and formatting contributor input for publication. In order to meet a strict deadline for project milestones or progress payments, publications had to be submitted on schedule.

Other rocket contracts consisted of providing project-manager support to the engine manufacturer for heat transfer, stress, and vibratory analysis of components. Individually executing such a work-scope—worth hundreds of thousands to millions of dollars on each of multiple engines—bordered on the impossible.

Integrating a team of engineers and analysts with requisite skills, each contributing to planning and execution of project deliverables, brought successful completion. Over the course of the program, contributing engineers were paid handsomely.

From my compensation, strategic investments went to three churches and multiple ministries working around the world. Reproduce that impact among a billion evangelical Christians around the world, and you have a formula for funding worldwide growth of the church and making disciples. Money lacks intrinsic substance; only currency (transactions) creates value. Money does not move God; believing God creates currency (circulation).

Major technical or engineering endeavors integrate a wide array of leaders, designers, and contributors to achieve something envisioned by one or more concept architects. Who receives credit for the resulting product? Everyone involved assumes some ownership of the most complex project arguing, for example, "I worked on that rocket engine."

The kingdom of God serves as a master plan for all major undertakings. Each participant supplies a seemly insignificant part to the overarching plan of God. Even the Chief Architect (God), who fits every component to the intended purpose, descended to earth and became the Chief Cornerstone (Jesus) who provided a sacrifice for us to accept as the unearned wages of eternal life.

In chapter four of Numbers, God instructed Moses on what the responsibilities would be for the Gershonites, Kohathites, and Levites in helping with the function of the tabernacle service. From age thirty to fifty, thousands of eligible men performed those roles, so each person's contribution constituted a limited role maintaining the successful operation of the tabernacle service.

More than implying we give ten percent back to God; His principle of tithe applies to how much you want him to multiply. What you ask Him can have great influence, so ask and believe for billions being given to facilitate the kingdom (and you're not taxed on what He does through many).

Believers doing the work of the kingdom now is crucial with Jesus returning soon as the Bridegroom of the church. Matthew 25:1-13 gives us a warning about the second coming of Christ with the soon-coming bridegroom in the parable of ten virgins. Like the story, Jesus will appear quickly after an announcement: "Behold, the bridegroom cometh; go ye out to meet him" (Matthew 25:6). The wise were prepared and entered, while the foolish who did not were excluded and told, "I know you not."

Something Fat and Sweet

Scripture contains a series of narratives tying together stories, books, and principles across the ages. One such theme covers the principle of blessing. Patriarch Abraham gathered his servants to pursue the armies of four kings who had attacked the cities in the valley of Siddim around the Salt Sea (Genesis 14:1-24), carrying away the inhabitants and spoils. Though living a herder's existence, dwelling in tents around the area of Canaan, Abraham managed to accumulate a vast herding operation, a large enterprise for its time.

When Abraham and 318 of his men returned in victory over a vastly superior army of Chedorlaomer King of Elam and the other kings, he brought back the stolen spoils of Sodom, Gomorrah and their allies.

Central to this conflict was Abraham's effort to rescue his nephew,

Lot, who had been taken captive after pitching his tent toward Sodom. Eventually, Lot and his family moved into Sodom. Rather than keep those rightly obtained spoils, Abraham returned the goods to the kings of Sodom and Gomorrah, but not before giving tithe to Melchizedek King of Salem, priest of the most-high God.

Plainly Melchizedek provides a portrayal of Jesus Christ who will also be king *and* priest (Hebrews 6:20-7:4). Melchizedek brought forth bread and wine while blessing Abraham. Of itself this passage would be interesting were it not for the recurring scriptural mention of something fat and sweet within the context of blessing. In the Scripture, a blessing always included mention of fat and sweet; for example, milk and honey, butter and honey, corn and wine, bread and wine, meat and honey, and oil and wine.

God promised Abraham a two-fold blessing in that he would be blessed and be a blessing to the nations (Genesis 12:1-3). The promise would have passed down from Abraham to his son Isaac, meaning the blessing flowed to Abraham's seed by his offsprings' faith (Romans 4:13) and that stretched through the generations, all the way to the ultimate blessing for God's people and the world: Jesus Christ.

Scripture says, "And to thy seed, which is Christ (Galatians 3:16)." Abraham's son Ishmael by his wife's servant Hagar was cast out with his mother, assuring that the blessing passed down to Isaac.

When it came time for Isaac to pass on the blessing, he selected the oldest son Esau, but Jacob (or deceiver) tricked both Esau and Isaac to gain the birthright (double portion) and blessing. Esau was described as a hunter and Jacob a plain man who stayed back at the tent. Naturally his father preferred the outdoorsman over a homebody (fathers want their sons to be the star, the football quarterback).

At the encouragement of Rebekah, Jacob proceeded to steal Esau's rightful blessing by fooling his blind father into thinking he was blessing Esau. Notice the expression of fat and sweet by Isaac in transmitting the prayer of blessing to Jacob:

> Therefore God give thee of the dew of heaven, and the fatness
> of the earth, and plenty of corn and wine: Let people serve thee,
> and nations bow down to thee: be lord over thy brethren, and
> let thy mother's sons bow down to thee: cursed be every one

that curseth thee, and blessed be he that blesseth thee (Genesis 27:28-29).

On detecting his brother Jacob's deception, Esau tearfully pleaded with Isaac and only received half-hearted encouragement, while assuring Esau that the blessing upon Jacob could not be reversed.

Jacob finally met his match in Laban. While plagued with adversity during his tenure working as a shepherd for his father-in-law, Laban, by his usual deceit, Jacob set aside a flock for himself. Given the mostly uniform color of Laban's flocks, Jacob attempted to induce the flocks and herds to reproduce with spotted and speckled coats by using notched and streaked rods.

Jacob's plan had appeared to work, but when a conflict arose with Laban, God spoke to Jacob in a dream, showing him how God had intervened on Jacob's behalf. Whatever course Jacob chose, God blessed him, so Jacob realized that God did not require his guile to succeed.

When the time came for Jacob (renamed "Israel" by God) to pass the birthright and blessings on to one of his twelve sons (Genesis 49), Jacob gave reasons why the older sons Reuben, Simeon, and Levi would be passed over for the blessing.

On his deathbed, Israel (or Jacob) bestowed a double-portion on Joseph (Israel's first son with Rachel), by making Joseph's sons Ephraim and Manasseh equal heirs with Joseph's brothers in the inheritance from Israel (Genesis 48:22).

Then Israel passed the blessing to Judah, the fourth oldest son. This all came after years earlier when ten of the brothers tried to convince their father Israel to send Benjamin with them to Egypt. Benjamin was Israel's only remaining son of Rachel (their last son overall) who was still with Israel after Rachel's first son Joseph had been sold by his brothers into slavery in Egypt.

Israel refused to send Benjamin, lest evil befall him like Joseph. To reassure their father of Benjamin's safety, Reuben offered to have his own sons slain if he did not bring Benjamin back. But then Judah made a personal sacrifice by offering himself as security, saying, "…then let me bear the blame for ever (Genesis 43:9)." From Judah's sacrifice, Israel pronounced the blessing by declaring something fat and sweet:

He washed his garments in wine, and his clothes in the blood of grapes: His eyes shall be red with wine and his teeth white with milk (Genesis 49:11,12).

Thus, Judah inherited Israel's blessing that had been passed down from Abraham to Isaac and then to Jacob (Israel). The offspring coveted the blessing more than the birthright, since the latter was a one-time double portion of the inheritance, while the blessing continued and affected every facet of life.

Something fat derives from the Hebrew for "oil" or "to shine" as the anointing which represents abundance. The concept of sweet reflects "joy" (Nehemiah 8:10). Biscuits need butter and honey!

For anyone who has experienced abundance, there is nothing so empty as having plenty but being without joy. Sports heroes, Hollywood idols, or popular recording artists can relate to having everything you could want but not enjoying it. The fulness of blessing includes both abundance of God's favor and joy in His Spirit.

One interesting picture of the blessing appears in the story of Samson (Judges 14:5-18) who went to Timnath to take a Philistine wife. Upon meeting a lion, the Spirit of God came upon Samson and he slew it with his bare hands.

Palestine air is quite arid, so carcasses dehydrate without decay. Samson later returned home on the same road and found a hive of bees with honeycomb inside the lion's carcass. Drawing a parable from this seeming paradox, Samson posed a riddle to the wedding party and challenged them to solve it: "Out of the eater came forth meat, and out of the strong came forth sweetness." (Judges 14:14)

Under duress from his Philistine bride, Delilah, Samson divulged the puzzle. She told the other Philistines at the wedding party, so they proudly announced: "What is sweeter than honey? And what is stronger than a lion?" (Judges 14:18)

Samson Eating Honey from a Lion Carcass

Jesus is referred to as the Lion of Judah (fourth son of Israel), the rightful heir of Israel's blessed birthright, which means Judah's lineage blessed the people of Israel and the world with Jesus. In Jesus's Last Supper, He shared the elements with His disciples and said:

Take, eat: this is my body, which is broken for you: this do in remembrance of me. After the same manner also he took the cup, when he had supped, saying, This cup is the new testament in my blood: this do ye, as oft as ye drink it, in remembrance of me (1 Corinthians 11:24,25).

In Jewish culture, the heir of the blessing need not share it with his brothers, but Jesus—the rightful beneficiary of the blessing—did not celebrate receiving the blessing himself at the Last Supper; instead, He demonstrated he would become the world's blessing by offering both His body and blood.

The fatness and sweetness, or abundance and joy of the blessing comes not only in receiving it, but also in doing as Jesus did in multiplying His blessing by dividing it and sharing it with others. The Hebrew word for Messiah reflects the "anointing with oil" or "to shine." A son (in Hebrew "ben" derived from "build") as heir of the blessing portrays the offspring as builder of his father's house.

A final concept to share on biblical blessings is found in Revelation. At the opening of the third seal with commencement of great famine, a voice among the four beasts commands: "…and see thou hurt not the oil and the wine (Revelation 6:6)."

During famine in Isaac's days, God warned him against going to Egypt (as his father Abraham had done), but instead to remain in the land, so Isaac took his herds and entourage to nearby Abimelech, king of the Philistines in Gerar (Genesis 26:1). At this same time, God reiterated the blessing Isaac had received, passed down from those originally promised to Abraham. As a result for Isaac while living in Gerar during famine:

> Isaac sowed in that land, and received in the same year an hundredfold: and the Lord blessed him. And the man waxed great, and went forward, and grew until he became very great (Genesis 26:12-13).

In time of greatest affliction, God prospers a blessing for the faithful; not a blessing of receiving or preserving, but multiplication through investing in others. At the coming of Christ prior to the Day of the Lord's wrath, those who take the blessing, break it, and multiply as Jesus did, receive much more as did faithful stewards in the parable of talents (Matthew 25:14-30). Jesus said, "…this do in remembrance of me" (Luke 22:19); emphasis fundamentally on "do" more than just "remembrance".

Value Creation

If you display a paper $20 bill to an audience and ask, "What is this worth?" there will be a few attempts at humor like "Not much," or "Less than last year!" But most affirm it's worth $20. Then for a sleight of hand, ask something like, "Would you exchange your hat for this $20?" If the answer is "Yes," the $20 suddenly can be said to be worth the hat—and vice versa.

That is the concept of currency. It is a value exchange or medium of exchange. At the time when gold and silver backed our currency (when the U.S. government would exchange dollars for precious metals), the dollar's value was backed by a commodity with intrinsic worth.

In fact our coins were made from metals bullion that included copper, silver, or gold that had worth. But value can only have meaning in the context of an exchange of products and services, whether or not a medium like money is used. Money or currency only has value when flowing or being exchanged. Unused, money has no intrinsic value; only an impression on metal or ink on paper.

In the classic Christmas movie *It's a Wonderful Life*, George Bailey was confronted by a run on his institution, Bailey Building & Loan, when a large group of his customers were not able to immediately access their funds.

Some pointed out that Mr. Potter, who purchased the Bedford Falls Bank, was offering to buy out Bailey Building & Loan and pay each of Bailey's customers only fifty cents on the dollar for their accounts, arguing that half was better than nothing.

George Bailey appealed to a mob of customers by protesting that they needed to understand what was happening. Mr. Potter was taking advantage of the bank-run crisis by offering to give the panicked customers their deposited money back, but at fifty percent off what the Building & Loan owed them. Potter was buying at a discount to increase his assets by taking over Bailey Building & Loan.

Moral concerns aside, the ability to buy when everyone is selling, characterizes those who know a classic formula of financial success: "buy low, sell high." In this case, rather than buy when the crowd is buying, Potter was attempting to find something undervalued and purchase it at that bargain price and hold it the value increased again.

Another investing metaphor involves sailing. Whether getting from point A to B or boating for pleasure, sailing skills provide lessons for life. First, certain influences can prevent a captain from achieving his objective if not using key principles like knowing the voyage's purpose (commerce, pleasure, transportation) and focusing on its objective, having the skills to operate the boat, and understanding navigation rules that avoid collision.

For a sailboat, navigation involves not only understanding boat operation, but also sensing the physics of wind and seas. Sailing with the wind still takes some skill, but a novice trying to move forward against a headwind can seem improbable.

How does someone make progress going into the wind? Well, rather than fight against it, you harness the wind. Going against a headwind requires using the sail and rudder to vector the wind, while sailing at an angle off-course from the destination. After a vector to the left, a vector to the right is needed, all the while keeping fixed on the objective. Tacking into the wind is slower than negotiating a tailwind, but it will get you to the objective. Following the wind is simple but won't get you to your goal if it's moving in a different direction and not harnessed skillfully with sail and rudder.

Bringing that sort of approach in sailing back to financial progress, crowds often fall into the trap of following trends. Making financial progress when prices are rising appears easy but does not increase value. Growth happens best in a gradual rise and fall by tacking back and forth (buying and selling) to make progress. Avoiding the temptation to follow the crowd by selling into a panic or buying in a bubble requires discipline just as sailing against headwinds. Know when to change course and tack the other way. Real, sustaining growth happens best with the gradual rising and falling of assets, like tacking back and forth into a headwind will also get you to the goal.

In the mid-nineteenth century, John Chapman (1774-1845) moved across the American Midwest planting apple orchards. He lived inconspicuously, possessing minimal provisions and collecting discarded apple seeds from cider mills.

The legacy of this man who came to be known as Johnny Appleseed was one of a missionary and orchardist, resting on his commitments to plant apple orchards while sharing his faith as he went across Pennsylvania, Ohio, Kentucky, and Indiana.

Beginning with limited resources, over his lifetime, Chapman accumulated 1,200 acres of apple orchards. Sadly, some of Chapman's views were idyllic, and failing to register his property, his heirs lost claim to much of it.

At the heart of our metaphors on investing, that every person moves through life by plan, chance, or some form of neglect. God created a course laid out to fulfil his plan, which, if followed results in only good, regardless of appearances on the surface. As Jesus stated:

> If any man will do his will, he shall know of the doctrine, whether it be of God, or whether I speak of myself (John 7:17).

Prophet for Profit

If you're a parent you know that we sometimes give children approval that depends on them doing something, only to see them take it as authorization. After mom says, "Only if your father approves," when dad is approached it becomes "Mom says it's okay." Likewise, a wide chasm separates what our will desires and what God has planned for us.

Only by drawing closer to God and aligning with His purpose can the error of Balaam be avoided. Balaam, son of Beor (Numbers 22), found himself with a potentially lucrative opportunity caused by Israel leaving Egypt to venture through the wilderness toward the Promised Land.

Having a reputation as a prophet who could successfully bless or curse nations, Balaam could usually name his price. But dwelling in the desert region of nations like the Moabites and Midianites did not come with many large paydays. So Balaam was initially excited when emissaries from Balak, the prince of Midian, came to Balaam with a profitable offer that was the stuff of dreams—a once in a lifetime windfall.

But then an insurmountable problem popped up: Balak wanted Balaam to curse the nation of Israel who had camped nearby, putting the people of Midian at risk. God wouldn't allow Balaam to curse God's people, and refused to allow him to go with Balak's emissaries. Disappointed, Balaam replied to Balak that he was on orders to not go beyond what God allowed, and he would not violate that rule. Was Balaam a true prophet, or just a soothsayer for hire?

Then Balak upped the offer by sending more prominent

representatives with a better deal. Still Balaam would not budge saying he could not go beyond the word of the Lord. Balak next sent his best offer, "…I will do whatsoever thou sayest unto me:…" (Numbers 22:17).

Though Balaam again responded in kind, this time he volunteered to ask the Lord again. Perchance something changed and he encouraged the visitors to tarry overnight. And wouldn't you know, God told Balaam he could go "if" the men came to call him. That was close enough. The next morning Balaam saddled his donkey and left with the men.

Balaam on His Donkey Confronted by an Angel

At this point God intervened by sending an angel to destroy Balaam. Great! You are doing what God wants and you get punished. Fortunately for him, the donkey three times recognized the threat, diverted course, and saved Balaam's life. And when Balaam disciplined the donkey, God spoke through the "dumb ass" (2 Peter 2:16) to rebuke him. Suddenly, God opened Balaam's eyes to see the angel and God Himself called Balaam's way perverse. But then God permitted him to proceed provided he strictly obey God's directives.

Where did Balaam go wrong? As with all of us, Balaam made up his mind he wanted to go and was pushing the envelope by asking God, even though he knew he shouldn't go. With the huge financial prize in his head, Balaam was looking for enough wiggle room to get his foot in the door. Though God's "if" was conditional, it was all Balaam needed to interpret it as approval.

So after Balaam received a rebuke through a donkey, God gave him consent to continue, but with the strict requirement to speak only what God authorized. Balaam complied without yielding to pressure from Balak despite three requests to curse Israel, instead blessing Israel three times while cursing Moab, Midian, and other nations.

At this point Balak gave up on the curse, yet Balaam proceeded (Numbers 24:15-19) to pronounce a prophetic assessment and acknowledgement of the coming "…Star out of Jacob, and Sceptre out of Israel…".

After all the fuss, Balak and Balaam returned to their places—but it doesn't end there. Israelite men began to commit sin with Moabite women, resulting in God's punishment of Israel. As later revealed, Balaam went on to counsel Balak that he could get vengeance against Israel simply by inviting them to sin (Numbers 31:16). If you are raising children, you want them to obey. If you are raising adults, you want them to learn to make good decisions of their own will.

Satan must always obey God's directives (Job 1), though absent any direct command and given opportunity, the devil seeks his own way, resisting the plan of God. Like Satan, the prophet Balaam was someone who put personal gain above integrity.

Subsequent references to Balaam in Scripture (2 Peter 2:15, Jude 1:11, and Revelation 2:14) picture him as self-willed, interested in his own gain, and resistant to the plan and purpose of God. Bible prophecies of what is still to come tell us the Antichrist and false prophet—as Satan's proxies on earth—will work at destroying God's faithful and for them to resist His plan. Jesus warned of these great troubles that will precede His return:

> And then shall many be offended, and shall betray one another, and shall hate one another. And many false prophets shall arise, and deceive many. And because iniquity shall abound, the love

of many shall wax cold. But he that shall endure unto the end, the same shall be saved (Matthew 24:10-13).

Jesus tells of an unjust judge who actually did the right thing, despite lacking fear of God or man. The judge accommodated a widow who incessantly pleaded for justice, and from that Jesus pronounced the lesson learned for those who love God:

Hear what the unjust judge saith. And shall not God avenge his own elect, which cry day and night unto him, though he bear long with them? I tell you that he will avenge them speedily. Nevertheless when the Son of Man cometh, shall he find faith on the earth? (Luke 18:6-8)

The King's Broad Arrow

Chapter Thirteen - Signs Ahead

E ARLY IN THE history of the American colonies, during the era of timber shipbuilding, and in the eastern-forests areas, the British government instituted a doctrine called the King's Broad Arrow of 1691. It was brought on by England's decline in suitable, tall, straight trees like the white pines found in the colony woodlands.

The policy confiscated all suitable trees for the Royal Navy using the doctrine evolved from the British government practice identifying property with a stylized arrow. Trees marked at the trunk with three axe cuts in the form of a vertical arrow identified the king's property and their theft (even on personal land) could result in heavy fines or imprisonment.

Resistance of colonists to the policy led to a Pine Tree Flag, symbolizing one of many revolutionary grievances. *The King's Broad Arrow* by Kathryn Goodwin Tone was a book that incorporated the theme and underlying influences.

Like the edicts of kings, often set boundaries can have severe and unreasonable consequences, as also happened in the day of Abraham. Twice while Abraham dwelled nomadically in Canaan, he and Sarah journeyed out of the wilderness to governed nations.

Early in their travels, they left their region in response to famine and dwelled for a time in Egypt under a pharaoh of the Middle Egyptian Kingdom (Genesis 12:10-20). Then past the age of sixty-five and still childless, Sarah gained a reputation in Pharaoh's court among the princes who saw her striking beauty and reported it to Pharaoh. Prior to entering the city of Pharoah and knowing Sarah would attract attention, Abraham advised his wife to refer to him as her brother (rightly so, since he was her half-brother).

Though already under the call and blessing of God (Genesis 12:1-3), Abraham's entourage and wealth had not yet grown to a conspicuous level. One day emissaries from the court suddenly arrived and took Sarah to Pharaoh's house, with the express purpose of adding her to his harem.

Normally, strategic alliances among chieftains and rulers included exchange of prized brides or concubines to cement peaceful cooperation. Leaders in Abraham's day regarded attractive women as the exclusive right of the ruler, and failure of a man to respect the custom by marrying a beautiful female was risking death.

That practice clearly drove Abraham's request of Sarah to act as if they were siblings, both with Pharaoh, and twenty-five-years later (Genesis 20:1-13), when a similar situation arose with Abimelech ("father of the king") the King of Gerar.

In the case of Pharoah, he quickly returned Sarah when God brought a plague on him, and despite Abraham having received a handsome reward when his "sister" was confiscated, Pharoah returned Sarah with no request for repayment.

A quarter century later, a similar interaction with King Abimelech resulted in much weightier consequences. God warned him in a dream of certain death for him and his household, after which he informed his servants to their great distress.

Twenty-five years earlier God had promised Abraham and Sarah a child without fulfillment. Abraham, now age ninety-nine and Sarah at eighty-nine, received a visitation from the Lord assuring the promised child when three men (or angels)—one of them the Lord—showed up at the door of their tent (Genesis 18:1-22). There the Lord spoke to Abraham of a promised son (Isaac, meaning "laughter") within the next nine months. Overhearing the prediction, Sarah laughed at the idea.

Following this, the two other angels descended into the cities of Sodom and Gomorrah to rescue Abraham's nephew Lot and his family, before the impending destruction. Shortly thereafter, Abraham, Sarah, and their company moved camp toward Gerar and an uncomfortable interaction with Abimelech.

Now recently with child by Abraham, Sarah entered the household of the king, risking public scandal should she be violated by Abimelech. As with the divine intervention on Pharaoh's intentions with Sarah twenty-five years earlier, God rebuked Abimelech and his house, at which point Abimelech declared (and God affirmed) that she had not been touched.

Sometime later, Abimelech and his chief minister, Phichol, visited Abraham's encampment at Beersheba, seeking a covenant peace agreement. As a goodwill sign, Abraham gave Abimelech sheep and oxen, as well as seven ewe lambs that were set off by themselves and given as a token of the oath. The place, which had a well, was named Beersheba ("well of oath," or "well of seven").

In Hebrew, *oath* and *seven* pertain to completion; thus, swearing an oath denotes completing oneself. God swears by Himself because no one and nothing greater exists to swear on behalf of:

> Look unto me, and be ye saved, all the ends of the earth: for I am God, and there is none else. I have sworn by myself, the word is gone out of my mouth in righteousness, and shall not return, That unto me every knee shall bow, every tongue shall swear (Isaiah 45:22,23).

The oath of seven at Beersheba lays a foundation for the future covenant of seven to come between the Antichrist, nations, and Israel, an agreement that Isaiah labeled a "covenant with death" and an "agreement with hell" (Isaiah 28:14-18). (We'll discuss the final covenant in more detail in Chapter seventeen.)

That last-days period of seven years is talked about as two periods of three-and-a-half years each. A current effort following from this ancient covenant at Beersheba can be seen with today's Abraham Accords being expanded to include more of Israel's adversaries.

Curiously, thirty years after the Beersheba covenant, Abraham's son Isaac made a similar covenant (Genesis 26:26-33) with a descendent of Abimelech, near the well Beersheba. The name Abimelech serves as both proper name and title. In both covenants King Abimelech recognized the blessing and anointing of God upon Abraham and Isaac.

Position or privilege can come from physical attractiveness, titles, intelligence, skill, or finances, whether obtained by birthright, effort, or chance. And God maintains providential rule over those and all things:

> In the day of prosperity be joyful, but in the day of adversity consider: God also hath set one over against the other, to the end that man should find nothing after him (Ecclesiastes 7:14).

As with all of God's designs for persons or people groups, Satan plots to subvert and resist His plans. But at every turn, God finds means to privately or publicly demonstrate how He alone controls men's future, the course of nations, and the reins of history.

Just as Abraham intervened to help his nephew Lot, God controls His calendar leading to the redemption of believers—so He alone will orchestrate the coming closure of our current age by calling out those followers at the sixth seal with cosmic signs, followed by the beginning of the Day of the Lord's Wrath, to be poured out upon the disobedient of this world.

Rest

In 1973 when Debby and I served with WMO as missionaries in Port au Prince, Haiti, we developed a friendship with another missionary family. As a skilled mechanic in another ministry, Roger (not his real name) provided assistance addressing maintenance issues with WMO's Land Rover and Jeep. During visits with Roger, he challenged some of my beliefs, particularly those related to observing the Sabbath. As a result of the issues he raised, it forced me to take a deep dive into the basis for my scriptural beliefs.

Roger's first target was about God's commandment that instituted the Sabbath and his second involved when the day of rest changed from Saturday (the Hebrew's seventh day of the week) to Sunday (the Hebrew's first day of the week). You've no doubt heard the expression that the Bible is a book of answers. But more importantly it's a book of right questions. Quoting Scripture does not necessarily supply answers without also asking God to interpret, understand, and put into context what you've read. Though it may seem a subtle distinction, the result is profound.

Over time the first century Christian Church transitioned from largely Jewish believers to Jews and Gentiles and then a myriad of Greeks and other non-Jews. Eventually even Jewish believers in Jesus found themselves excluded from the synagogue.

To prevent entrance, an oath against Jesus would often be demanded. Addressing the trend, Paul admonished, "No man speaking by the Spirit of God calleth Jesus accursed..." (1 Corinthians 12:3). Hence, for assembly and worship the Christians chose the first day (Sunday) rather than attending synagogue on the seventh day (Saturday).

Did assembly on the first day miss the intent of the commandment? Why did God rest on and bless the seventh day? Was he tired?

Thus the heavens and the earth were finished, and all the host of them. And on the seventh day God ended his work which he had made; and he rested on the seventh day from all his work which he had made. And God blessed the seventh day, and sanctified it: because that in it he had rested from all his work which God created and made (Genesis 2:1-3).

Scripture's Hebrew wording states that God finally "rested" (or "ceased") because creation was finished. The meaning describes completion. There are several Hebrew or Aramaic words for completion and the Greek language expresses completion by the word *teleios* ("complete" or "as intended").

Often Hebrew or Greek words for completion are translated as "perfect," for example in reference to Job's character. However, rather than "perfect," in this context "complete" more correctly means "as intended," rather than implying something as flawless.

As relates to creation, events of the first six days are framed as complete in as much as there remained nothing left to do. Hence, God rested on the seventh day and sanctified the day because He was finished.

The principle of a seventh day as rest can be seen across much of Bible history and in many cultures, being well established even prior to the Ten Commandments given in Exodus 20. Though not rigorously enforced during much of the Old Testament, this changed when the Israelites returned from exile at the time of Ezra and Nehemiah (560-430 BCE).

And by the New Testament days, Sabbath observance had become a pinnacle of the legalistic Sadducees and Pharisees. Though theologically liberal, Sadducees dominated the governing Sanhedrin of the time, while strict measures of enforcement were the hallmark of the more numerous and conservative Pharisees.

Before going further, consider the meaning of "complete" (or "as intended"). Engineering is often termed a teleological science. That is, focusing on purpose rather than cause. While science seeks to clarify fundamental laws governing a particular discipline, engineering utilizes those principles to design a device or system to accomplish a *purpose*.

When a device satisfies all the stated requirements and can be shown to fulfill the purpose for which it was conceived, a design may be termed complete, or as intended. New parts undergo First Article Inspection or an Acceptance Test to verify all requirements are met. That does not imply perfect in the sense of never requiring further modification. Frequently a new requirement, unforeseen condition, or failure develops, requiring a device or system to be redesigned.

> I have made the earth, and created man upon it: I, even my hands, have stretched out the heavens, and all their host have I commanded (Isaiah 45:12).

After the six-day creation when the heavens had been fully stretched out, everything was finished according to God's purpose. Even the Tree of the Knowledge of Good and Evil was as intended, which meant it was subject to God's constraints He put on it: "Thou shalt not eat of it: for in the day that thou eatest thereof thou shalt surely die" (Genesis 2:17).

God established something with conditions, and as we all know, constraints put on people introduce temptation, which does not help when a serpent's voice whispers, "Yea, hath God said, Ye shall not eat of every tree of the garden?" (Genesis 3:1)

Eve knew God's requirement and corrected the serpent about how they could eat from every tree except for only one. But then Eve added a condition God had not said: "…neither shall ye touch it, lest ye die." Where did that come from?

Because God spoke the initial restriction to Adam before Eve existed, Adam had to have warned Eve in his own words. Either Eve spent too much time admiring the tree, or from an abundance of caution, Adam likely advised to not even touch it.

Perhaps he should have added, don't look at the tree or go near it. Or better yet, Adam could have planted a hedge about the tree. Not that it mattered. As soon as Satan countered with,

> Ye shall not surely die: For God doth know that in the day ye

eat thereof, then your eyes shall be opened, and ye shall be as gods, knowing good and evil" (Genesis 3:4,5).

That sounds like a "friend" accusing your parents of preventing you from having fun. What's interesting about the tree is that Satan did not tempt Adam and Eve with some alternative he created, but rather used a lie about what God made.

Instead of life's challenges being addressed by learning of some alternative reality, the solution comes from understanding Satan's deception about what already exists. That deception is the limit of what Satan can offer.

Remember the lesson of Delilah who badgered Samson until he confessed that his uncut hair was the source of his great strength. Never tell anyone you have a secret you cannot share.

Once the serpent accused God of keeping something from Eve, her fall was sealed. Eve was going to check it out, notwithstanding the cost. To her, the consequence of death was a vague concept. So Satan introduced sin and death into the world, and with it, the fall rests on all mankind.

But God had a plan for our redemption foreshadowed by an animal sacrifice with the covering for Adam and Eve, as well as a promised seed to Eve who would bruise the head of the serpent (Genesis 3:15). With the prophesy, however, conditions are set to make the problem become part of the solution.

And the Son of man, Jesus, would become the final spiritual blow to the serpent. The Book of Hebrews describes the purpose of God's final sacrifice.

But we see Jesus, who was made a little lower than the angels for the suffering of death, crowned with glory and honour; that he by the grace of God should taste death for every man. For it became him, for whom are all things and by whom are all things, in bringing many sons unto glory, to make the captain of their salvation perfect through sufferings (Hebrews 2:9,10).

Why add suffering to make Jesus perfect? In the context of the previous translation for the Greek word *telios* ("complete" or "as intended"), Jesus as the Son of God, came into the world having not experienced the consequence of suffering as every person does. So Jesus was made complete or as intended by feeling the same emotions as every person who has ever lived. When bearing the sins of the world, he cried out: "My God, my God, why have you forsaken me?" (Mark 15:34). Jesus took the problem (my sin) and bearing it on the cross, made the problem part of the solution.

Of the Ten Commandments, one can be termed impossible. Each of the ten laws from God are not ends in themselves, but rather given to point out a higher purpose. They were provided to Moses because the children of Israel did not abide by the covenant with Abraham, Isaac, and Jacob.

That covenant of circumcision represented an inward relationship of the heart symbolized by outward manifestation. The Ten Commandments, just as civil law, do not exist as an end in itself. A society can succeed only if the populace restrains themselves apart from law, while law creates a behavioral perimeter.

Jesus said, "Ye have heard that it was said by them of old time, Thou shalt not kill..." (Matthew 5:21), while also noting that the intent is even stricter; do not hate.

Jesus also said, "Ye have heard that it was said by them of old time, Thou shalt not commit adultery... (Matthew 5:27)." The intent was to not lust; a much more rigorous law. Thus, the intent created a greater restriction than the commandment.

Each person must live within the scope of law to avoid violating it, which brings us back to, "Remember the sabbath day, to keep it holy." Original limitations noted in Exodus 20:8-11 applied to family, servants, and beasts not working on the seventh day (Saturday). These eventually were expanded to become rigorous legalism by the time of Jesus, so much so that the practice of keeping the original law and almost endless, added law missed the original intent.

Today within operating software, appliance manufacturers include a Sabbath Compliant Mode for Star-K Kosher certification that can be set to ensure food prepared on the Sabbath requires no programming work.

Yet someone must work to assure the power is generated and

supplied. So all the legalistic constraints make true rest on the Sabbath elusive, which is why it's the "impossible commandment". King David the psalmist declared:

> O come, let us worship and bow down: let us kneel before the Lord our maker. For he is our God; and we are the people of his pasture, and the sheep of his hand. Today if ye will hear his voice, harden not your heart, as in the provocation, and as in the day of temptation in the wilderness: when your fathers tempted me, proved me, and saw my work. Forty years long was I grieved with this generation, and said, It is a people that do err in their heart, and they have not known my ways: Unto whom I sware in my wrath that they should not enter into my rest (Psalm 95:6-11).

Because the people saw God's "work" in the wilderness but did not believe, they followed their own ways and could not enter His rest. For thousands of years throughout the Old Testament times, promises to Noah, Abraham, Moses, David, and the prophets spoke of a coming Redeemer Who would break the curse of sin. It culminated on Calvary with Jesus nailed to a cross, bearing the price of mankind's sin.

The gospels record Jesus saying seven things on the cross, and the sixth was, "It is finished" (John 19:30). Christ alone completed the work. With this declaration, the power of the curse was broken and rest restored to all who accept His salvation work as finishing any requirement for us to be reconciled to eternal relationship with God—if we'll just accept it. For those who believe, today constitutes our Sabbath rest (Hebrews 3:7–4:9). No additional work needs to be done. Working for your own salvation has the consequence of death.

Within this passage from Hebrews, the terms "today," "daily," and "day" are used eleven times to denote Sabbath rest as every day of the week—so it's today: Monday, Tuesday, Wednesday, Thursday, Friday, Saturday, and Sunday. In the Genesis 1 creation account, God alone worked, and he sanctified or set aside the seventh day because he rested after he finished his work.

Likewise on the cross, Jesus Christ finished His work of redemption for all who repent and believe. And on the coming Day of the Lord,

God alone will finish His work in this world by wrath which will be poured out on the ungodly.

Don Quixote Jousting with Windmills

Symbols Versus Substance

Originally published in 1605 by Miguel de Cervantes, the Spanish novel *Don Quixote* follows the travels and activities of aspiring-noble Alonso Quijano (Don Quixote) as he and his squire Sancho Penza seek to restore chivalry. His impractical adventures for social change descend into many ill-fated ideas, like attempting to joust a windmill he mistakes for a giant.

Don Quixote's exploits have become the metaphor for well-meaning-but-misaligned priorities.

Years ago, when our granddaughter Hannah's school class received an assignment to make a windmill from recycled materials, she created a wind-turbine model by repurposing hardware from a Habitat for Humanity store. Venetian blind louvers and a banister post served as the essential elements. Courtesy of my woodworking shop and her effort, the final product was a turbine able to orient to the wind and spin on the blade axis, having the appearance and functionality of a wind turbine, but short of producing useful work. Though only a symbol of the real thing, her project received first prize.

Wind Turbine Model

Once during one of my church training classes, a participant raised a complaint about Christmas trees, crosses, and flags as idolatry. Often Scripture employs a parable of images to portray ideas, issues, and spiritual principles. As Jesus taught, the parable is a riddle that must be unraveled. Particularly in the Book of Proverbs, Hebrew expressions fashion a paradox contrasting ideas where interpretation requires insight to understand the substance behind the symbolism.

Unfortunately, for much of culture and life, symbols get in the way of substance. Symbols themselves are not the problem; only when symbols are all we possess and become an end in themselves.

For those who embrace only symbols, a cross obscures the meaning of the sacrificial-death behind it, a church building replaces the body of Christ, plaques of the Ten Commandments are elevated over a relationship with God, and a dusty Bible substitutes for knowing Jesus as the living Word of God. Warfare for the believer does not involve destroying the symbols but should focus on the calling to make disciples.

Christian discipleship is simply disciples making disciples who make disciples. That principle is symbolized by God through His creation of self-replicating seeds, each reproducing after its kind.

When Christ returns to rule and reign, there will be no statues to pull down or other symbols to destroy: the anarchists will have finished that work. Satan fights against the symbols because he cannot influence or destroy the substance.

For believers, the plan and purpose of God will be fulfilled when Jesus returns at the sixth seal (Revelation 6:12), for the catching away of believers, to return again when Christ reigns His rightful kingdom from Mount Zion.

Pillars of Wisdom

Chapter Fourteen - Seven Pillars

FOR SEVERAL YEARS during our teens, my brother John and I ventured out of the house at night and went into town. Often it involved a prank around Halloween. Locking the bedroom door and with an alarm clock under the pillow, we would exit the window onto a porch roof and then jump ten feet to the ground.

Sneaking back into the house required creativity and various strategies that were always stressful from the risk of being caught, so one day we noticed a solution. Under our bedroom window, a large chest (with a closing seat cover) sat next to the enclosed porch we needed to access so we could reach the attic between the porch roof and ceiling from under our window.

A few measurements assured us that the chest's back could get us to the interior of the porch attic enclosure if we cut out a panel at the back of the chest. Our new trap door would lead into the porch ceiling and down into the porch. After a few saw cuts, we created a false panel and used a rope ladder for convenience. That became our way of discretely getting out for night-time adventures, while returning the same way an hour later.

One Halloween my mother's dress dummy served as an elaborate ruse. With a busy home, her relic stood in the attic for years until we rescued it from the cobwebs. My brother and I added artificial arms to the mechanical carcass, and a papier-mâché skull. Finally, a draped sheet transformed the dress dummy into a realistic, heart-stopping grim reaper.

Positioning the reaper in the woods, we would sneak out at night to create excitement along the rural highway. Around a sharp bend in the road, appearing from behind a giant oak tree, the sudden apparition of death personified in the headlights, guaranteed screeching tires and a terrified driver. The oak tree provided shelter to quickly grab the grim reaper dummy, drop it in a ditch, and run for cover.

With sudden disappearance of the grim reaper, even the stoutest-hearted driver would not venture into the woods to check out what he thought he saw. Imagining tales told at home by drivers that night served as entertainment enough. Looking back, likely a few of them did some soul searching after that night's confrontation.

Years later, Dad found our hidden, escape door and complained that the home's structural integrity must have been compromised. But we assured no structural elements were harmed in our making the trap door. Eventually we disclosed how it had given us late night access to and from the house.

Fortunately, for us and our neighbors we wisely outgrew the creative pranks, while no one was seriously injured in the process. The grim reaper project represented one of many outlandish youthful adventures, replaced later by more productive engineering work and construction of several homes.

The concepts of wisdom and truth always get a lot of attention, but apart from a transcendent, all-knowing, all-powerful God, there cannot be true meaning to wisdom. A verse from Proverbs embraces the interesting concept of wisdom being founded on seven pillars that are built upon a proper foundation, "Wisdom hath builded her house, she hath hewn out her seven pillars" (Proverbs 9:1).

From an engineering perspective on structural integrity, every building and element of it must maintain its proper load path. In Hebrew, the word *ben* (meaning son) derives from *banah* (to build), so Ben David (son of David) reflects a builder of the house of David.

As already mentioned, the biblical seven represents the concept of completion and being perfected—being finished. Some have attempted to list principles of wisdom, but the Word of God is the final authority on that:

> Who is a wise man and endued with knowledge among you? Let him shew out of a good conversation his works with meekness of wisdom. But if ye have bitter envying and strife in your hearts, glory not, and lie not against the truth. This wisdom descendeth not from above, but is earthly, sensual, devilish. For where envying and strife is, there is confusion and every evil work. But the wisdom that is from above is first pure, then peaceable, gentle, and easy to be intreated, full of mercy and good fruits, without partiality, and without hypocrisy. And the fruit of righteousness is sown in peace of them that make peace (James 3:13-18).

From this passage we get Scripture's seven pillars of wisdom:
1. Pure
2. Peaceable
3. Gentle
4. Easy to be entreated
5. Full of mercy and good fruits
6. Without partiality
7. Without hypocrisy

Contrast these seven pillars of wisdom with character flaws James points out in the same chapter: earthly, sensual, bitter envying, strife, and confusion, as well as every other evil work.

Consider all these contradictory qualities in relation to the parable of Ten Virgins (Matthew 25:1-13). As with the women's preparations (or lack thereof) prior to their potential bridegroom suddenly appearing, as the soon return of Christ approaches a preoccupation with normal activities of life will distract many. In Luke 17, Jesus spoke about Lot fleeing the destruction of Sodom and Gomorrah:

> They did eat, they drank, they bought, they sold, they planted, they builded. (Luke 17:28)

There is nothing inherently wrong with any of those activities. The problem arises from missing the work of God. In the process they also missed His rest, instead inheriting His wrath, as God warned:

> Your fathers tempted me, proved me, and saw my work. Forty years long was I grieved with this generation, and said, It is a people that do err in their heart, and they have not known my ways: Unto whom I sware in my wrath that they should not enter into my rest (Psalms 95:9-11).

Wisdom, Understanding, and Knowledge

Godly wisdom and His revelation are inseparable. The secular education system identifies constructs about knowledge development and use. What is knowledge? What makes a thing, statement, or idea true or false? How does understanding and wisdom fit in?

In general, the humanistic view of knowledge is that it is acquired by experience and training. Worldly understanding involves fitting knowledge into education and idea models before applying that knowledge to life. So in this worldly context, wisdom is acquired with time, study, and experience. But does that make sense from a biblical perspective?

Experience teaches that wisdom often does not come with age and more learning. We all know someone for whom this expression applies: "There is no fool like an old fool." A worldview approach claims that knowledge leads to understanding—which develops into wisdom. But this secular approach is like establishing a home by first buying furniture, then building walls around it, and finally trying to fit the house's foundation under them. In contrast, Scripture tells us the correct way:

Through wisdom is an house builded; and by understanding it is established: and by knowledge shall the chambers be filled with all precious and pleasant riches (Proverbs 24:3-4).

In Hebrew, wisdom is the starting point of construction where "to build" refers to "the foundation." Then by understanding the house is set up or established by erecting the walls. Finally, by knowledge it will be filled with riches or furnishings. Job spoke of God:

And unto man he said, Behold, the fear of the Lord, that is wisdom; and to depart from evil is understanding (Job 28:28).

In order to truly experience the grace of God and true wisdom, knowledge, and understanding, it must begin by opening the window to God's revelation of Himself. When we know the nature and person of God, He supernaturally reveals Himself. Then we can first understand issues and then the answers.

Knowing God is Not Understanding God

God created man and woman each as a reflection of His divine nature. But by themselves, both are incomplete. Though God speaks of Himself as "He" and Scripture addresses God's person as "He," there are references to female characteristics as relates to His nature. Note that God does not relate Himself to male and female; rather, male and female qualities by themselves give an incomplete image of God.

And even a male and female together in marriage are not complete without relationship to God. That marriage relationship is not finished until each person stands in the presence of God. Jesus said, "For in the resurrection they neither marry nor are given in marriage, but are as the angels of God in heaven" (Matthew 22:30).

We must know what God reveals of Himself to have a meaningful relationship with Him. Though we can know God, we cannot claim to understand God.

Contrast that with the believer's understanding of God's adversary. We do not know Satan, but we must understand his schemes. As the enemy of everything good and godly, Satan ("adversary" or "resist") finds every opportunity to resist God's will and purpose. Satan will seek to subvert and destroy everything that fulfills God's will, and when he cannot destroy God's truth, he'll seek to corrupt, counterfeit, and undermine it.

Where did Satan get his start? Scripture first introduces him as the cunning serpent in Genesis 3 with his temptation of Eve in the garden. Later, he appears in a variety of contexts, but always with a singular purpose: to resist God and His plan for mankind. Nowhere do Satan's character and conspiratorial plans show more clearly than Isaiah 14 and Ezekiel 28.

In Isaiah 14:12, he's identified by the Hebrew name *Heylel* (shining one or morning star). Notice that this is commonly translated in most Bible versions as the Latin name Lucifer which derives from Hebrew *halal* ("to shine" "praise" or "act foolishly").

The Hebrew word *halal* serves as a root of the Hebrew expression *hallelujah* ("praise to Jah," or "Jehovah").

The term Lucifer ("light bearer") derives from the Latin Septuagint (Greek) translation rendering of the Hebrew archangel name Heylel.

His role included the architect of worship and music in the presence of God (Ezekiel 28:11-19). Embellished with wisdom and beauty, Heylel elevated himself in pride above God, precipitating a rebellion and his downfall. Of three archangels Michael, Gabriel, and Heylel, the last ended up resisting God and leading a third of the angels to resist with him. About Heylel, God said:

> Thou art the anointed cherub that covereth: and I have set thee so: thou wast upon the holy mountain of God; thou hast walked up and down in the midst of the stones of fire. Thou wast perfect in thy ways from the day that thou wast created, till iniquity was found in thee (Ezekiel 28:14-15).

A classic example of Satan's resistance movement against God can be seen in the history lesson of Stephen, the first Christian martyr. When called before the Sanhedrin, the foreign-born and Hellenistic-Jew Stephen was accused by false witnesses of blasphemous statements against the temple and law. The high priest demanded Stephen answer, "Are these things so?" (Acts 7:1)

In response, Stephen proceeded to identify each of the historical instances where their ancestors resisted God's messenger in the persons of Joseph, Moses, the prophets, and finally Jesus himself. Then he went on to critique the Sanhedrin, "Ye stiffnecked and uncircumcised in heart and ears, ye do always resist the Holy Ghost: as your fathers did, so do ye" (Acts 7:51).

How ironic that Stephen had to give these scholars the history of Israel's leaders who consistently attacked God's plan as it was revealed. Irony is the essence of humor when catching the listener off guard by a sudden reversal of expectations, just as Christ will surprise when He interrupts today's world on its path to self-destruction.

Absolute Truth

A few absolute truths:

- Jesus Christ is the Son of God.
- God created all things by His Word.
- Jesus Christ came in the flesh, was crucified, died, and was raised again to life.
- God is omniscient, omnipresent, and omnipotent.
- Jesus Christ exists eternally.
- Jesus Christ will return to rule and reign on earth.

Contrary to popular claims, these truths do not originate from simple faith, tradition, history, doctrinal position, or personal conviction. Such objections are demonstrably false. Those who repent of their sins and have faith in the finished work of Christ on the cross to cover those sins will experience an understanding of Him that includes forgiveness of sin, knowledge about Jesus, and the revelation that a relationship with Christ is the only way to God. These truths are revealed in the Old and New Testaments and have been affirmed by believers throughout the history of the church.

Particularly outstanding are the testimonies of Christ's former enemies like Paul the Apostle, who came to know Christ after persecuting believers. Paul testified to knowing truths that were revealed to him directly from God.

Lee Strobel, author of *The Case for Christ,* serves as a contemporary example of an antagonist lawyer and former editor of *The Chicago Tribune* who, when challenged to investigate the claims of Christ's resurrection, found them to be indisputable.

Jesus, Judas, and the Purse

Chapter Fifteen – Crossover

AT WORK SOME years ago, three-pound cannisters of coffee were disappearing from a break-room cabinet. After purchasing bulk supplies, the coffee club noticed many cans missing. This theft took place several times and could not have been successful without somehow hiding the cans while going out of the building entrance, so we reported the issue to management and security.

Following up with security after a few weeks, we were assured, "It is being handled." A short time later an employee vanished unannounced; a shameful career sacrifice for a few dollars of coffee.

If you have worked in a group with discretionary monies, you probably know that misappropriations are not uncommon. Eventually someone notices the missing funds and the culprit is usually someone with access to the funds. No one likes a thief, so rumors quickly get around and word filters up to the one in charge.

Scripture does not record whether the disciples raised their concerns with Jesus about Judas stealing from their common purse—not that

Christ needed to be told. What's curious is that Jesus never openly relieved Judas of handling the ministry's funds. However, Jesus likely put him on notice to "give account of thy stewardship" (Luke 16:2). Judas may have still possessed the purse at the Last Supper (John 13:29), but something changed for Judas a few weeks before the crucifixion. Was he given his two-week notice and told to give account of the purse's money?

John comments (John 12:5,6) that Judas—bearer of the common purse among Jesus and the twelve—was a thief. John infers that the disciples must have known Judas was skimming from the bag for personal use. Either they saw it happening or surmised his deception because he purchased things the other disciples could not.

Near the conclusion of Jesus' ministry, seemingly liberal-minded Judas accused Mary of wasting costly ointment on Jesus, which could have been sold for much and given to the poor. Both Matthew (26:14-16) and Mark (14:10,11) record how Judas' decision to betray Jesus happened immediately after the confrontation about Mary using all that expensive oil to anoint Christ's feet.

In this light, examine the Luke 16 story of the unjust money manager (steward). Considering the surrounding events from the gospels and the timeline of Jesus' ministry, He told this story in the final two weeks before the Last Supper. He said that a master accused his financial manager of misappropriating funds and told the thief to turn in an accounting of his work before being let go.

Hoping to gain the goodwill of his master's clients and thus find a new job after leaving, the manager decided to curry favor with those debtors by wrongfully reducing their balances owed to the master.

Surprisingly, when learning of the manager's scheme, rather than venting further outrage at the crooked steward, the master commended him for his cunning. In the lesson to this parable, Jesus observed:

...for the children of this world are in their generation wiser that the children of light (Luke 16:8).

What prompted Jesus to tell the story? Why would a master commend his manager (steward) for being cunningly unfaithful? The answers reside in the story of Judas. Why did Judas betray Jesus?

Certainly, there were a number of factors motivating Judas, but in large part it seems to have been because the ministry was beginning to diverge from Judas' expectations.

Whether or not Jesus directly called Judas to account for stealing, the story in Luke 16 likely hit home with Judas. Seeing the implication as a warning of soon dismissal and disgrace, it probably prompted him to quickly take action to protect his future. The solution would be to curry favor with the Jewish leadership who hated Jesus, wanted Him gone, and had the resources to reward Judas. What better way than to give them what they wanted; for a price, say thirty pieces of silver (the current price of a common slave).

When Stephen was tried for his faith and delivered a long, historical discourse to the ruling council of Jerusalem, he accused their Jewish ancestors of betraying their brother Joseph by selling him to the Midianite merchants for twenty pieces of silver. Painting the current leaders of Israel with the same treasonous brush, Stephen declared before being stoned to death:

> Ye stiffnecked and uncircumcised in heart and ears, ye do always resist the Holy Ghost: as your fathers did, so do ye (Acts 7:51).

Joseph was sold into Egypt for twenty silver pieces foreshadowing Judas' betrayal of Jesus for thirty pieces of silver. Why only twenty for Joseph? Traveling merchants headed to Egypt had no need of slaves, so to them he only had value as resale merchandise on the Egyptian slave market. The best Joseph's brothers could do with him was to negotiate a wholesale price of twenty pieces of silver.

Intertwined stories like these demonstrate complex threads tying together the whole of Scripture. And throughout the Word of God, Satan enters, always seizing opportunity through his minions to resist and betray the plan of God.

But in the final chapter to come just before the rule and reign of Christ on earth for a thousand years, God will vanquish all of Satan's plans of resistance. Stephen cast the leaders of Israel's Sanhedrin (religious leaders) in the same mold as Satan, whose ultimate plan and purpose stands against what God wills and purposes for man. The enemy's entire agenda will fail on the coming Day of the Lord, when only the work of God will prosper.

Potiphar's Wife

If given lemons, make lemonade. That sage, paraphrased wisdom is mentioned in Dale Carnegie's book *How to Stop Worrying and Start Living*. Favored by God and their father, Joseph fell on the wrong side of his brothers' affections.

Of course, he had no one to blame but himself with his unwise self-promotion in reporting his dreams, and relaying his brothers' misbehavior. More importantly for Jacob's entire family, God had a purpose for the all of the troubles in their midst.

One day when Joseph visited his herdsmen half-brothers tending the flocks, the ten of them concocted a plan to kill Joseph and resolve their source of family conflict. After Reuben and Judah objected to the plan, the brothers compromised, agreeing to "only" sell Joseph into slavery with some passing Midianites.

So Joseph was taken to Egypt where he was resold into slavery under Potiphar, the captain of the guard. Joseph turned out to be insightful and skilled at financial management, prospering and blessing Potiphar's household and other affairs.

Joseph with Potiphar and His Wife

Also favored by God to be handsome, Joseph caught the eye of Potiphar's wife, and after a few failed attempts at seducing him, she cornered him long enough for a proposition. At that point Joseph again refused and fled, leaving his robe in the hands of Potiphar's wife.

In order to protect herself, Potiphar's wife falsely accused Joseph of assaulting her. As a result, Potiphar imprisoned Joseph. But God had a higher purpose for Joseph. Did Potiphar see through his wife's accusation? Did he later change his thinking about Joseph?

Remember that Joseph and Potiphar had prospered as well as the household. And then with Joseph thrown into prison, the prison keeper saw Joseph's value and favored him with work as the de facto prison manager.

Where specifically was Joseph imprisoned? As it turns out, Potiphar was not only captain of the guard but also held oversight of the king's prison and its inmates (Genesis 39:20). How do we know? Later when the king's butler and baker were jailed (Genesis 40:3), Pharaoh had them put into the prison in Potiphar's house.

This means Potiphar provided a service to Pharaoh by maintaining a detention facility or dungeon in his home. In fact, the captain of the guard (Potiphar) assigned the new prisoners (the butler and baker) to Joseph (Genesis 40:4).

It seems Potiphar sensed Joseph's innocence, but it was not prudent to release him without implicating his wife for slander of Joseph. Sometimes the course of least resistance appears lower risk.

We learn that all of Joseph's affliction had been working for his good, given the opportunity to interpret the butler and baker's dreams. As a result, two years later, Joseph received a call to interpret Pharoah's dream after all of Egypt's wise men failed.

Suddenly, Joseph's gift of interpretation earned him a promotion from prisoner to Pharaoh's prime minister, second in command over all of Egypt. And who then began reporting to Joseph? His former owner Potiphar! By this time Potiphar had certainly put two and two together!

As with Joseph's outcome, here on the doorstep of the last-days of history, God will soon intervene suddenly to turn all things for the good for those who have been trusting in Him. At Christ's appearing in the sky, both the dead in Christ and alive will be caught up to be with Christ before God's wrath is poured out on earth and all those remaining. Just as Joseph was quickly promoted from prison to prime minister, Christ's followers will be suddenly be raised to rule and reign under Him for a thousand years.

The Churl is Not Liberal

The apparent concern of Judas for the poor certainly displayed his liberality. What does it mean to be liberal? Is liberality merely a contrast to conservative? Notice the first warning of Jesus on the Mount of Olives when teaching of His return He cautioned, "Take heed that no man deceive you." (Matthew 24:4) In an effort to clarify terms, the modern meaning of liberality in educational, political, economic, or social issues does not reflect the biblical position. Isaiah prophesied:

> The vile person shall be no more called liberal, nor the churl [an impolite or mean-spirited person] said to be bountiful. For the vile person will speak villainy, and his heart will work iniquity, to practice hypocrisy, and to utter error against the Lord, to make empty the soul of the hungry, and he will cause the drink of the thirsty to fail. The instruments also of the churl are evil: he deviseth wicked devices to destroy the poor with lying words, even when the needy speaketh right. But the liberal deviseth liberal things; and by liberal things shall he stand." (Isaiah 32:5-8)

True, biblical liberality refers to someone with a generous and selfless spirit. The perfect example of liberality was Jesus at the Last Supper, offering himself for the good of the world and telling his disciples, "... do this in remembrance of me." (Luke 22:19) The emphasis should be placed on "do" more than on "remembrance".

Contrast the Bible's liberality with Isaiah's churl who is "a rude, stingy, or morose person." So don't mistake true, scriptural liberality with the modern culture demonstrating churlishness as liberality.

Under the inspiration of the Holy Spirit, Paul the Apostle properly expressed what people would be like in these times:

> This know also, that in the last days perilous times shall come. For men shall be lovers of their own selves, covetous, boasters, proud, blasphemers, disobedient to parents, unthankful, unholy, without natural affection, trucebreakers, false accusers, incontinent, fierce, despisers of those that are good, traitors, heady, highminded, lovers of pleasures more than lovers of God; having a form of godliness, but denying the power thereof: from such turn away (2 Timothy 3:1-5).

Spoiled On Purpose

God instructed Isaiah to marry a prophetess, bear a son and call him Mahershalalhashbaz. When God assigns a name, it carries a message. In this case, the longest name in Scripture comes with the banner, "hasten the spoil, speed the captivity" where *shalal* means spoil. To the cities of Damascus (Syria) and Samaria (Israel) this word testified to the coming spoil of nations and people set against God. With the coming of Assyrian armies, Syria and Israel were destroyed and spoiled. Judah, under leadership of Hezekiah, escaped destruction until one hundred and fifty years later at the hands of Nebuchadnezzar and the Babylonians.

In the story of Ruth, Boaz the kinsman redeemer, said to his reapers,

> "...let fall also some of the handfuls of purpose for her, and leave them, that she may glean them, ..." (Ruth 2:16).

Ruth and Boaz

The law specified fallen grain to be the provision of poor and widows. During the harvest, if grain touched the ground it was spoiled or *shalal* (meaning plunder or spoil), and left for gleaners. In the story of Boaz redeeming the widow of a near kinsman, Jesus as redeemer, allowed himself to fall to the ground by a willful sacrifice of Himself to save all who believe in Him.

Boaz as redeemer directed the reapers to make a willful sacrifice just as Jesus directed his disciples at the last supper to "do this in remembrance of me". As Jesus said,

> "Verily, verily, I say unto you, Except a corn of wheat fall into the ground and die, it abideth alone: but if it die, it bringeth forth much fruit. He that loveth his life shall lose it; and he that hateth his life in the world shall keep it unto life eternal." (John 12:24).

Accursed

God charged Joshua as they approached Jericho with the complete destruction of the city and everything in it. All the spoil of Jericho was appointed by God to be accursed or *cherem* (meaning devoted or accursed). In Leviticus, the Lord instructed Moses,

"None devoted, which shall be devoted of men, shall be redeemed; but shall surely be put to death." (Leviticus 27:29)

Again, the word for devoted or *cherem* refers to the concept of something accursed or assigned to destruction, being devoted to God. Things devoted or accursed in this sense are assigned no useful means, apart from God's purpose. The message of the cross, like *shalal* and *cherem* presents something of intrinsic worth purposely assigned to destruction to bring forth greater value.

It sounds noble to be characterized as useful; particularly as it relates to the kingdom of God. An individual or a group of people may appear to have much to offer. That thinking misses the foundation of the kingdom of God. He does not need anyone, any class of people, or anything we have to offer. He completes himself, as I AM THAT I AM (*Hayah asher Hayah*). It is only by his mercy that God calls us to participate by faith with him in the kingdom of God that we might be rewarded by fulfilling his will. God himself completes the work of his kingdom. He does not require his followers to do the work, neither does he share the glory (Isaiah 42:8), though we are co-laborers with him.

The Cross Over

At commencement of the tenth plague, God announced to Moses,

"...I will pass through the land of Egypt this night, and will smite all the firstborn in the land of Egypt, ..." (Hebrews 12:12).

The word used here for pass through or *abar* (meaning pass over or cross over) can be tied intimately to the call to cross over by striking the blood on the side posts and upper post of the door. The corresponding term used at the Passover is *pasach* referring to the angel of the Lord passing over households with the blood on the door posts in the sign of a cross.

The expression *abar* meaning put away can be seen in Nathan's confrontation of David after ordering the killing of Uriah the Hitite, one of his leading men and husband of Bathsheba. When David pronounced to Nathan after his parable of the lamb, "...the man that hath done this thing shall surely die." (2 Samuel 12:5). Nathan responded, "...The Lord also hath put away thy sin; thou shalt not die." (2 Samuel 12:13)

The term put away or *abar* (to cross over) again images the cross as David had his sin covered over after repentance. The cross represents much more than a Christian symbol; it epitomizes an instrument of execution, or what Jesus accomplished near Mount Moriah.

David's response to personal conviction for his sin is recorded in Psalm 51 and also Psalm 22 where David's lament, "My God, my God, why hast thou forsaken me?", was repeated by Jesus on the cross. God crossed over David's sin, but there were grave consequences as Nathan pronounced the "sword will never depart from your house".

Costs to David included an assault on his daughter Tamar by oldest son Amnon, death of Amnon at the hand of Absalom for the assault, and rebellion and death of Absalom as well as the death of many in Israel.

Despite King David's anointing and favor, privilege took root when he refrained from participating in an ongoing war with the Ammonites (2 Samuel 11). Some background to David's situation will be helpful. A wise counsellor, Ahithophel participated among David's closest associates. Ahithophel's wisdom could be taken as "the oracle of God". Among David's trusted warriors, Eliam and Uriah (2 Samuel 23:34-39) participated in the chosen leadership. As it transpired, Eliam had a beautiful daughter, Bathsheba, who just so happened to be married to Uriah, the other elite commander of David.

Interestingly, Eliam's father Ahithophel was therefore grandfather to Bathsheba and great-grandfather of the child born, who eventually died for David's sin. In the mind of Ahithophel, he likely felt betrayed by King David for committing adultery with his granddaughter.

Also, Eliam's son-in-law Uriah was killed at the direction of David. That caused plenty of ill will to go around. Yet God chose to select David's next son Solomon by Bathsheba as heir to his throne. As such, Solomon son of David represents a type of Christ. God redeems all things, even failure. It is recorded that Nathan named Solomon Jedediah meaning beloved of *YAH* (2 Samuel 12:25).

John the Baptist and disciples of Jesus reverted to Bethabara at the crossing near Jordan, the preferred location for baptism. Throughout Scripture, the imagery of Jordan or *Yarden* meaning to go down or descend represents lowering to pass over as a transition. Just as Jesus descended to experience the condition of his creation and the cross, he returns descending by bending down and rending the heavens at his return to call believers out to himself.

Chapter Sixteen - First Coming

S ERIOUS AND OBJECTIVE investigation of biblical revelation attests to unique circumstances of Jesus Christ's lineage, prophetic fulfillment, and divine nature. As the Apostle John said,

" ...every spirit that confesseth not that Jesus Christ is come in the flesh is not of God: and this is that spirit of antichrist, whereof ye have heard that it should come; and even now already is it in the world" (1 John 4:3).

Not only did Jesus come into the world an amazing teacher and prophet, but his death, resurrection, and bodily ascension are without parallel. Add to these exceptional events, the unparalleled historical collaboration of fulfilled prophesy. The purpose of the gospel message testifies, not just that Jesus did something historically significant, but rather, as God incarnate, He redeems all who believe and receive him by faith.

History is filled with testimonies of changed lives as a result of knowing the risen Christ. The gospel message can be incapsulated as a personal revelation of Jesus Christ Himself to the heart, acknowledgment of the atonement for sin, and assurance of eternal life. Four experiences of relationship with God;

1. Personal knowledge of Jesus Christ as the Son of God
2. Knowledge of forgiveness of sin
3. Recognition that Jesus Christ is the only way to know God
4. Assurance of eternal life

Jesus Christ also came to fulfill the prophetic promise of redemption and spoke of ultimately returning to bring all things to completion and rest according to God's kingdom plan. He came as the promised Messiah of the Jewish people according to God's covenant with Abraham, as well as now to all peoples.

Nations have been blessed by embracing this redemptive message and others impoverished by turning from it. Strangely, in the purpose of God, the descendants of Abraham have been among the most stubborn adversaries of the gospel message, despite inheriting the favor of God. Of particular interest is how the Jewish people and the nation of Israel have been restored and have prospered after almost two thousand years of being scattered across the globe.

For example, consider the amazing statistics of Nobel Laureates. Between 1901 and 2020, there have been over 900 Nobel award recipients with 208 of them being of Jewish descent.

For a people group representing 0.25% of the world population, 23% of Nobel recipients are of Jewish lineage. That is almost a 100:1 ratio disparity. Ashkenazi Jews (descendants of exile) in particular, are disproportionally represented in academia, business, finance, entertainment, media, engineering, science, medicine, law and government.

Just as Joseph prospered in Potiphar's service, God uniquely placed the Jewish people in positions of influence around the world. When Joseph became second in command in Egypt and revealed himself to his brothers, he warned them to announce they were shepherds. Thereby he assured they would be despised by Egyptians and remain culturally separate in the land of Goshen.

Growing up in Western Pennsylvania during the 1950s-1960s provided insight into how pervasive antisemitic views become entrenched. Our family occasionally shopped at a nearby clothier run by Jewish owners. After mentioning a discount offered to my father by the proprietor to "friends of his family", my father dismissed, "That's just something Jewish."

Similar comments were voiced by classmates in high school. So it was enlightening the first year at Penn State University when a Jewish roommate from Philadelphia was assigned for the fall trimester.

Having met Adam (not his real name) during the summer trimester, it was unnerving to recall many of his proclamations and disdain for non-Jews. Adam, extremely outspoken, lacked any filter on his views. Notable also was the vast collection of LP phonograph albums stacked over his cabinets.

With great relief, he never showed up for fall trimester classes. As it turned out, his acquaintances said Penn State expelled Adam for misappropriating the property of others the preceding summer, particularly LP albums in the dormitory.

Now for the rest of the story. A few weeks into the fall session, a knock at the dorm room door revealed a new roommate assignee. Enter Jacob (not his real name). No greeting. No eye contact. Just pronouncements; he was here and not happy, except to finally get a dorm room after temporary housing.

Jacob spoke when absolutely necessary. Just the facts. No conversation, only a declaration of his issues. But then something transformational took place at meal time in the cafeteria. Seated around a half dozen or more complete strangers at the dining table, Jacob opened up, introduced himself, and within ten minutes befriended everyone with an easy-going banter.

It turned out he also hailed from Philadelphia, and he knew everyone and everything it seemed. Then, as suddenly as the conversation started, it degenerated into sarcastic comments to others at the table. Another ten minutes of back-and-forth and we are into "*airing of grievances*" a-la Seinfeld's Festivus. At that point Jacob stood up, made some pointed insults about non-Jews and left.

Rather than mend fences, the identical pattern repeated itself every meal. New strangers. Microwave friends. A mob scene. And a departure with name calling and insults. A short time later, noticing Jacob had conspicuously covered his middle name with tape on the door nametag, curiosity disclosed it to be "Israel." Another Jewish roommate from Philadelphia!

Once Jacob's background became well known, he made no secret of his disdain for anything Christian. The doctrine of the virgin birth took a particularly relentless beating. A friend of German descent down the hall became a target of particular scorn as "Kaiser Wilhelm."

After several months, a complaint to the Penn State housing office yielded only, "You need to learn to get along with all kinds of people." Over the course of three trimesters the issues did not improve, only softened, because by Christmas, Jacob declared he would no longer speak, and largely did not for the remaining five months.

For a number of years, our daughter Christina worked as office manager for a Christian ministry of Jewish believers with outreach to the Jewish community and the nation of Israel. These points of influence serve as seed sown to reap the harvest of the world in these end times.

As declared to the angel, "…Thrust in thy sickle, and reap: for the time is come for thee to reap; for the harvest of the earth is ripe." (Revelation 14:15)

As painful as the process at the university became, learning the importance of Jewish influence, their unique contribution to the world, and the purpose of God gives redeeming value to the story of why Jesus came to redeem a wayward creation.

My subsequent fifty-five years of ministry and career included numerous opportunities to work with and learn from Jewish people, and about Israel and the Hebrew language. Many of the included lessons are an outgrowth of studying the Hebrew language and the root meaning of Hebrew words.

Housing arrangements remained much calmer the sophomore through senior year at Penn State. The next assigned roommate, Ron, also from Philadelphia, turned out to be an ideal alignment of interests and temperament.

We spent many hours discussing our respective studies. Ron was doing cutting edge microbiology and DNA research, and mine were propulsion and space. A regular jest surrounded who might first win a Nobel prize in our field. After retirement from microbiology research, Ron launched into fabrication of small HO scale railroad miniatures. HO scale is the most popular model railroad size. Ron has created spectacular themes using scratch-built techniques to model the 1958 North Philadelphia community, railroads and industry.

The first coming of Jesus Christ descended through the lineage of King David. His ministry was initially centered around fellow descendants of Abraham, Isaac, and Jacob. In His crucifixion, Jesus extended His sacrifice to all the world.

God's calling always includes being in the world, but separate from the world. Despite calling as descendants of Abraham, most heirs of the blessing do not acknowledge Jesus Christ as the promised Messiah. In God's grand scheme, the redemptive message has been extended to the whole world.

Within recent decades the gospel message has been success-fully preached to all the nations and peoples of the world (Matthew 24:14), just as Jesus promised. To fulfill the covenant of God to Adam, Abraham, Isaac, Jacob, Moses and the prophets, He promised again to reveal Himself in the last days to the Jews and the nation of Israel (Isaiah 2:1-5). His return and the Day of the Lord approach quickly.

Chapter Seventeen - Second Coming

TRAVELING WEST ACROSS the expansive American prairie, dark shadows slowly dominate the horizon and quickly blossom into foothills, then a challenging mountain range: the Rockies. Signs forewarn, "Steep Slope Ahead," "Watch out for Falling Rocks," and "Impassible in Winter." There are few alternatives around the challenging ascent ahead.

Preceding chapters established context, traveling through forested lowlands with a few anecdotes and biblical principles. Ahead looms Mount Difficulty similar to John Bunyan's allegory in the story *Pilgrim's Progress*. Going will be slow, with narrow rocky roads well marked despite possible diversions.

As woodlands give way to spectacular scenic views along the route, the slope ahead rises quickly, and the rarified atmosphere can steal your breath away. But take heart. Once at the peak, the other side inclines leisurely to the conclusion and rest. For the faint of heart, you can circumvent the climb and go to the end, but there are a multitude of interesting observations ahead. After leading this trek the past thirty-five years, most who endure the ascent don't regret the challenge once on the other side.

The scene has been set for God's majestic and historical production of the last days. Prophetically, Jesus *will* return; not if. Much speculative ink has been spilled in anticipation of events surrounding Christ's second coming. When Jesus alluded to the destruction of the temple in Matthew 24:1,2, His disciples came to Him privately saying,

> Tell us when shall these things be? And what shall be the sign of thy coming, and of the end of the world? (Matthew 24:3).

Jesus began by warning of the risk for deception (Matthew 24:4,5). He then stated the popular pronouncement, "And ye shall hear of wars and rumours of wars…" (Matthew 24:6). Jesus followed the same verse with, "…see that ye be not troubled: for all these things must come to pass, but the end is not yet."

Of themselves wars and rumors of wars transcend history, but they are not an indication of the imminent return of Jesus. The corresponding passage in Luke 21 provides a clear indication of transition, saying:

> Then said he unto them, Nation shall rise against nation, and kingdom against kingdom: and great earthquakes shall be in divers places, and famines, and pestilences; and fearful sights and great signs shall there be from heaven. (Luke 21:10, 11)

The Greek for nation is the word *ethos* which speaks to ethnic divisions rather than geographic constraints. No explanation can adequately portray the rising tide of contemporary tribalism along ethnic lines. The problem of ethnic conflict tears apart families, nations, and the world. No greater challenge exists than that of hearing the loudest voices call out a problem "out there" and failing to look in the mirror and realize it is "in here." A personal problem lies at the root of the general problem.

Beginning Birth Pangs

What is tribulation? The Hebrew word *tsarah* for tribulation appears extensively in the Old Testament (Genesis 42:21) and means trouble or distress. Likewise, the Greek word *thlipsis* meaning tribulation reflects similar affliction. Jesus spoke in the synoptic gospel accounts of trials and persecution arising first from ethnic conflict, famines, pestilences, and earthquakes as the beginnings of sorrows (or birth pangs).

These stand as a forewarning of things to come (Matthew 24:8). Although similar conditions exist from place to place in the world and time to time through history, these events noted in the Olivet Discourse signal a dramatic, global intensification.

Tribulation has a global and historical footprint, not merely the concluding seven-year period called in Hebrew *shabua* meaning seven. Though ongoing suffering and distress and a time of great tribulation occur within this seven-year covenant, it should never be falsely labeled "the tribulation."

Deception

In like manner, the history of Old Testament and New Testament periods, as well as world cultures demonstrate a propensity to embrace false teaching, either for personal, professional, financial, social, or familial reasons. Sadly, many Christians do not avail themselves of information to evaluate and make sound biblical judgments regarding doctrinal issues.

Jesus alluded to the matter in a sequence of verses leading up to the middle of the seven-year period (Matthew 24:9-14). Discernment requires good offense, knowing the objective and embracing the truth, as well as defense through understanding and confronting the assaults of the enemy.

When an exceptional product dominates market share, competitors offer multiple cloned products, each of which capture incremental shares of the market. As a consequence, even a dominant product loses significant share to many inferior competitors.

The success of the gospel message fosters many alternative narratives, each with a different slant on three concepts of truth: the nature of God, the nature of man, and the means of reconciliation between God and man. Often differences in practice cloud the doctrinal scene and create unnecessary distinctives. Knowing the source of truth constitutes the best antidote to error. Satan excels at mimicking, not originality.

Roll with Seven Seals

Opening of the Seals

As with much prophetic revelation, the Apostle John stepped beyond the worldly existence of space and time, being commanded to, "Write the things which thou hast seen, and the things which are, and the things which shall be thereafter" (Revelation 1:19).

After the Lord's instruction to the seven churches, John beheld a door opened in heaven (Revelation 4:1-4), and heard a voice as a trumpet calling him to come up and be shown things to come.

In the spirit, John saw a throne and one that sat on the throne in heaven, in addition to twenty-four elders upon their seats and four beasts. At the presentation of a scroll with seven seals and a Lamb slain, the beasts and elders appear with a multitude of angels worshiping the Lamb. In response, all creation in heaven, in earth, under the earth, and in the seas bless the One on the throne and the Lamb.

At this stage the Lamb commences to open the seals. Of seven seals on the roll (book) opened by Jesus Christ the Lamb, focus here concentrates on the fifth and sixth seals. The timing for the commencement of opening the first seal is not recorded in relation to beginning the seven-year *shabua* period (84 months, 2520 days on the Jewish calendar based on a 360-day prophetic year). However, the fifth seal certainly marks the mid-point of the seven-years associated with the Abomination of Desolation. Following are the seven seals.

1. White Horse – Rider given a bow and crown, going forth to conquer
2. Red Horse – Rider given power and sword to kill and take peace from the earth
3. Black Horse – Rider with a balance, calls for famine, but ordered, "hurt not the oil and the wine"
4. Pale Horse – Death and Hell with power over a fourth part of the earth to kill
5. Slain Saints – Saints slain for the Word of God given white robes pending the deaths of fellow servants
6. Cosmic Signs, Heavens Depart – Catching away and the harbinger of the coming Day of the Lord's Wrath

   Sealing of 144,000, Great Multitude in Heaven—Remnant in Sela sealed in their foreheads
7. Coming Seven Trumpets – Silence in heaven, the prayers of the saints, and the preparation of seven trumpets

Of the seven, the first four seal judgments distress the whole earth, with no outward singling out of believers. Though these trials appear to reflect intensifying affliction, they are not solely directed toward Christians and Jews.

The first seal reflects conquering by strength, the second by war, the third famine following war, the fourth death following famine. The fifth seal parallels the Jewish Holocaust carried out by Antiochus IV Epiphanes at the Abomination setup in the temple as described in the Books of 1, 2 Maccabees.

These are also recorded in the twenty volume writings *The Antiquities of the Jews* by Flavius Josephus in 94 CE. As noted elsewhere, the caution, "hurt not the oil and the wine" (Revelation 6:6) alludes to God's protection concerning the blessing experienced by patriarchs, prophets, and saints during a time of great dearth.

Death and Hell represented by the pale horse at the fourth seal denote power over a fourth part of the land. The wording does not necessarily imply a fourth of the population are to be killed, but rather death will stalk a quarter of the earths' territory to kill by sword, famine, disease, and beasts. Use of the Greek word Hades (meaning departed) for Hell comes from Greek *a-eido* or unseen.

The fifth seal, reflected in three verses (Revelation 6:9-11) describes the impending death of Christians and faithful Jews prior to the sixth seal cosmic events and the catching away of believers. Beginning with the Abomination of Desolation, a period of Great Tribulation addresses the wrath of Antichrist against the church and Jewish people.

Of this period of months, Jesus said no flesh would survive were it not cut short prematurely (Matthew 24:22). After the sealing of the one-hundred-forty-four thousand Jews and the appearance of the great multitude in heaven, the seventh seal opens the seven trumpets judgment of the Day of the Lord's Wrath.

As explained in this writing, a major reset of space and time and the natural order occurs at the sixth seal when at the rolling away of the fabric of the heavens, remaining believers and the dead in faith rise and are caught away before impending wrath termed The Day of the Lord. Explanation of the sixth seal cosmic events follows later in this chapter.

Adonikim and Adonijah

Upon return from Israel's diaspora (scattering about) in Babylon and Persia, several groups were commissioned by Cyrus, Darius, and Artaxerxes to lead the Jews to Palestine (538-445 BCE).

Both Ezra and Nehemiah recorded the names of key family leaders and their respective households by the number of their descendants. In the tradition of the Book of Numbers, God declared that Moses number the names:

> Take ye the sum of all the congregation of the children of Israel, after their families, by the house of their fathers, with the number of their names, every male by their polls; from twenty years old and upward, all that are able to go forth to war in Israel... (Numbers 1:2-3).

Both post-exilic writers Ezra and Nehemiah record the names of family patriarchs and their respective numbers saying:

> ...The number of the men of the people of Israel... (Ezra 2:2)

One individual among the names has significance based on the Hebrew meaning *Adonikam* which is "Lord of rising" or "Lord of resurrection." The name Adonikam portrays one who resurrects or rises up. The following lists of chief men with numbers of descendants are not inclusive of all names given in Ezra 2:3-35. However, the name Adonikam in each listing appears in essentially the same order of the following chiefs as:

	Ezra 2	Ezra 8	Nehemiah 7	Nehemiah 10
9	Bebai (623)	Bebai	Bebai (628)	Azgad
10	Azgad (1222)	Azgad	Azgad (2322)	Bebai
11	Adonikam (666)	Adonikam	Adonikam (667)	Adonijah
12	Bigvai (2056)	Bigvai	Bigvai (2067)	Bigvai
13	Adin (454)		Adin (655)	Adin

The use of Adonijah in Nehemiah 10 rather than Adonikam will be discussed later in this chapter. The numbering sequence reflects a list of twenty chief names in Ezra 2 with only a portion shown for brevity. A basis for difference in descendant counts from Ezra 2 and Nehemiah 7 is not explained, but likely Adonikam (666) vs. Adonikam (667) reflects inclusion of the chief father himself rather than just descendants.

Nehemiah states (Nehemiah 7:5), "…I found a register of the genealogy of them which came up at the first, and found written therein." Since the numbers in Ezra and Nehemiah for other chiefs are not identical, it is evident the records do not reflect an exact copy. They were apparently made at different times for specific purposes, possibly by separate scribes.

An interesting connection in the two numbers associated with the name Adonikam (666) and (667) relates to the significance of numbers six and seven. The number six is associated with man and work as fallen and incomplete. Seven, however, in Hebrew means completion or perfection associated with God and His rest.

As developed following, the name Adonikam reflects someone who rises up. The rise of the Antichrist serves as Satan's final chance to usurp the crown of Jesus Christ who alone conquered death. By raising up a fallen man to rule the world, Satan will conspire to prevent the fulfillment of Christ's kingdom on earth. This beast, introduced in Revelation 13, refers to a head of the beast wounded to death and healed, causing the world to wonder.

While names and numbers by themselves do not necessarily carry significance in their own right, there is a connection with meaning tied to other references in Scripture. Specifically, the Lord revealed:

Here is wisdom. Let him that hath understanding count the number of the beast: for it is the number of a man; and his number is Six hundred threescore and six. (Revelation 13:18)

The association of name and number has no numerological meaning, but rather represents the count or number of persons associated with a prominent individual. Scripture explains Scripture. Unsuccessful attempts at use of numerology have given contradictory outcomes trying to derive 666 from letters of historical or present-day names.

At the unveiling of the Antichrist, unbelievers will identify with a mark, the name, or the number associated with the name (Revelation 13:17,18) in order to control commerce. The use of tattoos with the number, often combined with images of a pentagram and Satan, are already contemporary symbols of rebellion against authority. That is only a foreshadowing of something more sinister to come.

Biblical accounts of the Antichrist (Daniel 11:21-23) express how, though vile, he will gain rule by flattery, make a league with many, and work deceitfully to become strong with a small people.

In Isaiah (28:14-18), the prophet speaks to Israel about their "covenant with death" and "agreement with hell." This agreement, mentioned by the angel to Daniel (Daniel 9:27), reflects a truce with the nations guaranteeing the security of Israel subject to certain concessions.

The Hebrew word *shabua* for week or seven refers to the seventy sevens of years as well as to the final seven of years (Daniel 9:27). Other descriptions for the latter half of a final seven label the period as 42 months or 1260 days.

The covenant period, patterned after Abraham's covenant with Abimelech at Beersheba (well of the oath, well of seven) is outlined in Chapter 13 of this book as a framework for the final covenant of seven developing now as the Abraham Accords. The covenant is discussed further near the end of this chapter. This prince to come confirms a covenant with many for seven (Daniel 9:27), though he may not be the originator of the agreement.

The nation Israel insists on a guarantee of security which only God Himself can provide. What allegedly is for their protection becomes Israel's destruction when the Antichrist turns, in the middle of the seven-year agreement, to destroy Israel by violating the guarantee of protection (Daniel 11:30-31). The resulting Abomination of Desolation, spoken of by Jesus (Mark 13:14) can be seen by type and shadow through the working of Antiochus IV Epiphanes (Greek – resolute protector manifest) in the books of 1 and 2 Maccabees.

Two books of the Maccabean wars are included among seven

Apocryphal books of the Catholic Bible. Though not considered part of the canon of Scripture, they are a valuable historical record of the inter-testament period, which is 400 BCE (Before Christian Era) to 1 CE (Christian Era).

During the Protestant Reformation, Martin Luther translated a German Bible (1534 CE) from the Hebrew and Greek Septuagint, which included apocryphal books. After the death of Martin Luther (1546 CE), Catholic church leadership at the 19th ecumenical Council of Trent adopted apocryphal books from the Latin Vulgate into their bible as deuterocanonical writings.

Apocryphal books are not part of the Hebrew Scriptures nor the Protestant Bible because of their questionable author, origin and lack of divine inspiration. Events in Maccabees are also detailed in *The Antiquities of the Jews* by Flavius Josephus, a Jewish military leader who became historian to Emperor Vespasian after the destruction of Jerusalem in 70 CE.

Antiochus and The Abomination of Desolation

The Jews, Israel, and the church have variously experienced periods of persecution and tribulation including martyrdom since the first century. Most of the persecution remained localized and of varying intensity.

The Jewish Holocaust during World War II was one of the most severe periods of genocide in recent memory with an estimated death toll of six million Jews. By comparison, Stalinist era purges and pogroms are estimated to have killed about twenty million persons—for political, ethnic, and religious reasons. Descriptions of the horrors during these atrocities are appalling but few can compare to the documented exploits of Antiochus IV Epiphanes against the Jews as described in 1 and 2 Maccabees.

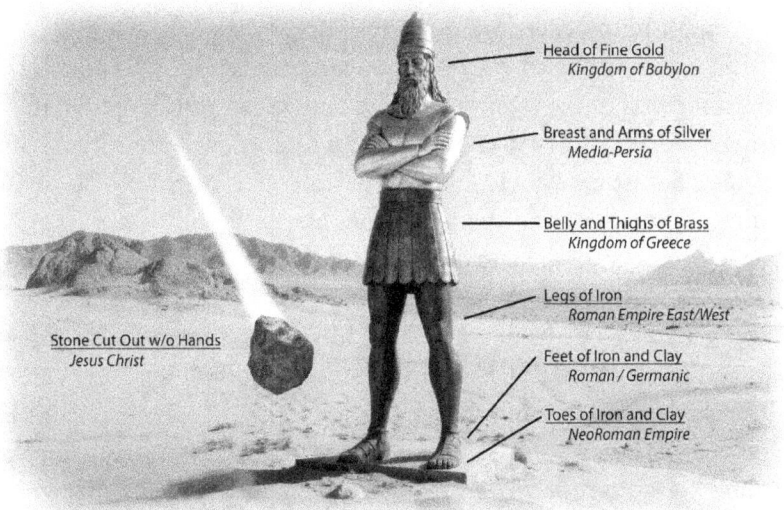

Head of Fine Gold
Kingdom of Babylon

Breast and Arms of Silver
Media-Persia

Belly and Thighs of Brass
Kingdom of Greece

Legs of Iron
Roman Empire East/West

Feet of Iron and Clay
Roman / Germanic

Toes of Iron and Clay
NeoRoman Empire

Stone Cut Out w/o Hands
Jesus Christ

Statue of Nebuchadnezzar's Dream

The Prophet Daniel recorded several dreams and visions regarding a succession of nations ruling Palestine including Babylon, Media-Persia, Greece, and Rome, as well as a collaborative successor to Rome. Identification of these kingdoms aligned with the nations described by the successive metals in the statue of King Nebuchadnezzar's dream. Each successive kingdom and ruler experienced some aspiration of world domination. To some extent all were successful in measure.

Of particular interest was the meteoric rise of Greek influence under Philip II of Macedon. Philip II, a successful Macedonian military strategist and his wife, Olympias, a devotee to the serpent-handling cult of Dionysos, gave birth to a son, Alexander. His birth coincided with a number of favorable omens.

Olympias, determined to elevate her son Alexander to the throne at all costs, convinced Alexander that he was not fathered by Philip but rather the Greek God Zeus. As a young military commander, Alexander soon demonstrated military prowess by conquering most of the civilized world. But when Alexander the Great (356-323 BCE) died suddenly at the age of 32, his world-wide kingdom under Greek rule was regionally divided among four generals; Ptolemy ruled in

Egypt, Seleucius in Mesopotamia/Central Asia, Attalid in Anatolia, and Antigonid in Macedon.

These events were revealed to Daniel in a vision (Daniel 7-11) with understanding given by the angel Gabriel (Daniel 8:16 and Daniel 9:21). History provides names and events identifying kings of the Hellenistic Seleucids, kings that reigned over the Middle East, including Jerusalem and the Jewish people.

The dynasty of Seleucid kings ruled Palestine and surrounding nations until the rise of Roman power under Julius Caesar. At the pinnacle of Seleucid rule was Antiochus IV Epiphanes, who initially tolerated Jewish traditions but grew to despise the Jews.

Antiochus ruled from 175 BCE (Before Christian Era) until his agonizing death in 164 BCE. During his rule of Jewish lands, he decided to consolidate power by embracing Hellenistic Jews who adopted Greek traditions.

In order to quell an uprising among faithful Jewish believers, Antiochus outlawed observances and ordered all Jews to worship Zeus. At this provocation, Judah Maccabee led a popular resistance against the Seleucid army. Despite numerous defeats and suffering at the hands of the Greeks, the Maccabees eventually overthrew Antiochus and restored traditional worship to Palestine. Before his death, Antiochus lamented persecuting the Jews (2 Maccabees 9:12).

Of paramount significance during this uprising was the event referred to as the Abomination of Desolation or Holocaust. After King Antiochus decreed the abolition of Jewish customs and worship, he committed an act of conspicuous desecration against the altar and temple to defile it and prevent further use.

The king ordered a swine sacrificed on the holy place of the temple and erected a statue of the Greek God Zeus, having his own image substituted for the face of Zeus. With the temple defiled, Jewish worship ceased for 3-1/2 years (1260 days) until completion of the successful revolt. Cleansing and restoring the temple took an additional 75 days, coinciding with the 1335 days recorded prophetically in Daniel 12:12.

The miracle of the Hannukah lamp holy oil occurred when a day's supply lasted eight days, instituting the post-exilic festival of

Hannukah or the Festival of Lights. This occurred 75 days after the Day of Atonement (Yom Kippur) and is so celebrated to this day.

Did the followers of Jesus and the multitude of Jews understand the significance of reference to the Abomination of Desolation? Indeed, they were well documented in the writings of 1 and 2 Maccabees as well as Jewish oral tradition.

Today the Jews know and understand the meaning and significance of the World War II Holocaust. So also, in the time of Jesus the bloody purge of King Antiochus to defile the temple in 175 BCE (Before Christian Era) was fresh in their understanding. These trials of the Maccabean revolt, though outrageous in their own right, pale in comparison with coming persecution under the rule of the Antichrist. Jews, hearing of the Abomination, are warned to flee to the wilderness (Sela). Only one hundred forty-four thousand comply and escape in the middle of the seventieth seven.

Sela

During the Olivet discourse, Jesus forewarned believers of a coming Abomination of Desolation. As Jesus spoke,

> When ye therefore shall see the Abomination of Desolation, spoken of by Daniel the prophet, stand in the holy place, (whoso readeth, let him understand:) Then let them which be in Judaea flee into the mountains: ...for then shall be great tribulation, such as was not since the beginning of the world to this time, no, nor ever shall be. And except those days should be shortened, there should no flesh be saved: but for the elect's sake those days shall be shortened. (Matthew 24:15,16,21,22)

Flee where? Jesus and John the Baptist, during their respective ministries, retired beyond the mountain wilderness of Judea southeast of Jerusalem. Beyond this wilderness the landscape descends rapidly to Jericho and then the Jordan River valley, roughly 1400 feet below sea level.

East of the Jordan (meaning to descend or go down) the topography rises again to a great expanse of mountainous Jordanian desert. There in Jordan lies a region formerly belonging to Moab, Edom, and Ammon. In the wilderness of Moab, the former city of Petra remains nestled in a secluded valley.

City of Petra (Sela) in Jordan

Fifteen hundred years before the ministry of Jesus, the Lord called Joshua (meaning *YHWH* is salvation) to lead Israel over Jordan into the land of promise. In the Hebrew, the names Jesus (*Yeshua*), Joshua, Hosea, and Isaiah all have the same root, origin, and meaning.

Moses, striking the rock a second time to bring forth water (Numbers 20:11), was admonished for failing to speak as instructed. As a consequence, Moses, a figure of the law, died on the other side of Jordan, relinquishing his role to Joshua.

The meaning? Law does not lead to the land of promise, only salvation by God's grace. At the crossing of Jordan (Hebrew *abar* meaning cross over), Joshua instructed the priests to step into Jordan carrying the ark.

Joshua and Israel Crossing the Jordan Near Jericho

Upon entering the waters the Jordan's flow stopped, parted, and rose upon a heap. At the instruction of Joshua, a representative of the twelve tribes selected and carried a stone from Jordan to set up a sign to future generations in Gilgal. This process of descending, parting, and crossing over symbolizes events yet to come at Christ's return in the heavens.

Petra, owing to unique structures carved into the stone walls, has become a modern tourist attraction. Once home to wealthy Nabataeans traders, who developed a thriving community with private and public buildings sculpted from sandstone cliffs, the city prospered on a lucrative trading route of the Roman Empire.

Gradually Petra came under the control of Rome, giving way to a thriving Christian church after the rise of Emperor Constantine. In the latter years of the Roman Empire, Petra's influence waned as a place of commerce, culminating with a major earthquake in 363 CE (Christian Era). From then until the 1800s, Petra remained effectively unknown except to a small local population of Bedouin inhabitants. Today Petra receives tourist visitors from Israel and Jordon subject to the local political climate.

The Prophet Isaiah (Isaiah 16:1-5) spoke of this region as Sela, an early name for Petra, calling for outcasts fleeing the spoiler to dwell with

Moab until the indignation passes. As Daniel prophesied, notwithstanding the Antichrist's domination of world governments; Edom, Moab, and Ammon would escape out of his hand (Daniel 11:41).

The Lord also revealed (Revelation 12) imagery of the woman (a type of Israel) bringing forth a man child (a type of Christ). The woman escapes the serpent's destruction, fleeing to the wilderness (Revelation 12:12-17) in the middle of the seven years for a time, times, and half a time (3-1/2 years, 42 months, 1260 days).

Through this time of Great Tribulation, pervasive in the earth, one hundred forty-four thousand as yet unredeemed Jews steal away out of Israel to the wilderness, hidden from the Antichrist.

Isaiah (Isaiah 10:21-23) and the Apostle Paul (Romans 11:5) raise the specter of a small remnant of Jews who escape, sequester, and are redeemed at the appearing of Jesus Christ in person.

This event fulfills the prophetic promise of preparing, separating, and then sealing of one hundred forty-four thousand from twelve tribes prophesied in Revelation 7 prior to the opening of the seventh seal and the beginning the Day of the Lord's Wrath.

The redemption of the one hundred forty-four thousand (or remnant) at the end of seven years (7 x 360 = 2520 days) brings to completion a literal fulfillment of the Day of Atonement (Yom Kippur). For the church of true believers, a unique day approaches within the latter 3-1/2 year window of the seventieth seven or *shabua*. That unknown day and hour, comes suddenly, as the announcement of a new moon on the festival Rosh Hashana, proclaimed by the sound of a trumpet, upon sighting of the new crescent moon.

Michael the Restrainer

The Canon of Scripture records three chief angels or archangels by name; Michael (meaning who is as God), Gabriel (meaning warrior of God), and Heylel (meaning shining one or Lucifer). Both Gabriel (Daniel 9:21, Luke 1:26) and Michael (Daniel 12:1, Revelation 12:7) receive multiple references and appear in Old and New Testament accounts.

However, the Hebrew language name Heylel has only one Old Testament mention (Isaiah 14:12). Most Christian are unaware of the name Heylel because of the use of Lucifer from the Latin Septuagint translation. Notably, of the three archangels, Heylel's name is the only one of the archangels whose name references himself rather than God. Jerome, one of the early church fathers, substituted the Latin name Lucifer (morning star, shining one) rather than the Hebrew name. Other names of angels based on non-canonical literature are of uncertain origin and not addressed here.

Descriptions for this third archangel address him as the facilitator of worship, a proud pompous figure, and self-willed. In Isaiah 14, Heylel boasts "I will" five times:

> For thou hast said in thine heart, I will ascend into heaven, I will exalt my throne above the stars of God: I will sit also upon the mount of the congregation, in the sides of the north: I will ascend above the heights of the clouds; I will be like the most High. Yet thou shalt be brought down to hell, to the sides of the pit. (Isaiah 14:13-15)

Isaiah 14 and Ezekiel 28 reference the fall of this third archangel employing the titles King of Babylon and Prince of Tyre respectively. These two chapters lay out his fall from grace for pride and leading a third of the angels in rebellion against God. Most portrayals of this third archangel cast him in his fallen role as Satan (meaning adversary or accuser), Devil (meaning slanderer), and dragon.

All that see the fallen archangel Heylel at his appearing will be astonished saying,

> They that see thee shall narrowly look upon thee, and consider thee, saying, Is this the man that made the earth to tremble, that did shake kingdoms; That made the world as a wilderness, and destroyed the cities thereof; that opened not the house of his prisoners? (Isaiah 14:16, 17)

In Daniel 10, the prophet speaks of three weeks abstention from pleasant bread, flesh, and wine while he sought the meaning of a vision. During this time, a vision appeared to Daniel saying:

Now I am come to make thee understand what shall befall thy people in the latter days: for yet the vision is for many days. (Daniel 10:14)

And:

But I will shew thee that which is noted in the scripture of truth: and there is none that holdeth with me in these things, but Michael your prince. (Daniel 10:21)

Here, the Hebrew word *chazaq* (meaning to strengthen or restrain) reflects Michael's role as restrainer of evil against the people Israel.

Daniel 11 develops the prophetic foreshadowing of Antiochus IV Epiphanes and his campaign against the Jews during the period of Maccabean rule. As the narrative transitions from Greek rule (about 168 BCE) to the latter days in Daniel 11:36, Michael enters the picture again as protector of Israel. However, at this time Michael takes a different stance from his normal role as defender:

And at that time shall Michael stand up, the great prince which standeth for the children of thy people: and there shall be a time of trouble, such as never was since there was a nation even to that same time: and at that time thy people shall be delivered, every one that shall be found written in the book. (Daniel 12:1)

In this verse, the Hebrew word for stand up *amad* (meaning stand or stand still) implies Michael stands aside from his role as restrainer and permits the enemies of the Jews to attempt the destruction of Israel. God Himself through the action of Michael guarantees the protection of the nation and the people of Israel.

As a consequence of this covenant with death, Israel will sign under duress with her arch-enemy, an agreement guaranteeing the protection only God can give. Upon Israel's apostasy, God gives them, "what you demanded." The Antichrist then violates the agreement by turning to destroy the nation and people.

In the final days of King Saul's rule, he sought unsuccessfully to find and destroy David, the anointed heir apparent to the throne. Though given two opportunities to destroy King Saul, David would not raise his hand against the king. Instead, David waited for God Himself to reciprocate and judge Saul in the proper time.

Speaking through Nathan the prophet (2 Samuel 7:11-17, 1 Chronicles 17:7-15), God denied David's plan to build God a house, promising instead to build David a house founded upon his son.

That promise saw fulfillment first in David's son Solomon, and finally in Jesus, the Son of David. One purpose of the Day of the Lord's Wrath remains to bring the remnant to faith in the true Messiah of the Jewish people.

The final day of Yom Kippur brings in the remnant of the Jews and closes the kingdom of God. Satan, the embodiment of the fallen angel Heylel, wants nothing more than to delay this inevitable outcome as it signals the end of his attempt to usurp the crown and rule over earth, displacing Christ's coming Kingdom of David.

The Beast

History and Scripture are replete with foreshadowing of the final antagonist character, who like the Antichrist, comes out of obscurity and quickly usurps the authority belonging to another. A few of these historical figures in their time were thought by the church to be the Antichrist. In retrospect, these were only a prefigure of the person called the Little Horn (Daniel 7:8).

1. Antiochus IV Epiphanes, (216-164 BCE)
2. Emperor Nero, (37–68 CE)
3. Emperor Diocletian, (244-316 CE)
4. Emperor Napoleon Bonaparte, (1769-1840 CE)
5. Adolf Hitler, (1889-1945 CE)

One characteristic description of the Beast is;

…one of his heads as it were wounded to death; and his deadly wound was healed: and all the world wondered after the beast. (Rev 13:3).

As Jesus Christ was raised from the dead, so the imagery of the name Adonikam reflects rising up. It shows he will rise from a deadly wound according to the meaning of his name "Lord of Resurrection" or "The Lord Rises Up."

And what of the use of Adonijah (Nehemiah 10:16) instead of Adonikam? Occasionally alternative names are used or assigned as in the case of Pharaoh Necho giving Eliakim the name Jehoiakim (2 Kings 23:34).

Here is an interesting connection. Prior to the promotion of Solomon to the throne, Adonijah (meaning my lord is Jehovah) sought to elevate himself and usurp the kingdom (1 Kings 1:8). Though pardoned by Solomon for a time, a subsequent request by Adonijah to marry Abishag was interpreted as seeking the crown, resulting in Adonijah's' execution for treason.

This scenario with names and imagery of Adonikam and Adonijah provides a foreshadowing of the rebellion of Satan and the Antichrist. He must experience a deadly wound and be healed. Then Satan, through his resurrected minion the Antichrist, will attempt to usurp Christ's rightful crown and dominion in the final days before the catching away, and eventually be thwarted by the wrath of God.

In the coming peaceable reign of Christ, when he returns to rule for a thousand years, Satan must be bound and the Antichrist (or beast) cast into the lake of fire with the false prophet (Revelation 19:20). At the same time Christ slays the army following the Antichrist with the sword (Word of God) out of his mouth (Revelation 19:21).

However, after the thousand-year reign, Satan must be loosed for a final rebellion among unbelieving nations. Those unregenerate have their lives prolonged during the millennial reign of Christ as declared in Daniel 7:

> I considered the horns, and, behold, there came up among them another little horn, before whom there were three of the first horns plucked up by the roots: and, behold, in this horn were eyes like the eyes of man, and a mouth speaking great things. I beheld till the thrones were cast down, and the Ancient of days did sit, whose garment was white as snow, and the hair of his head like the pure wool: his throne was like the fiery flame, and

his wheels as burning fire. A fiery stream issued and came forth from before him: thousand thousands ministered unto him, and ten thousand times ten thousand stood before him: the judgment was set, and the books were opened. I beheld then because of the voice of the great words which the horn spake: I beheld even till the beast was slain, and his body destroyed, and given to the burning flame. As concerning the rest of the beasts, they had their dominion taken away: yet their lives were prolonged for a season and time. I saw in the night visions, and, behold, one like the Son of man came with the clouds of heaven, and came to the Ancient of days, and they brought him near before him. And there was given him dominion, and glory, and a kingdom, that all people, nations, and languages, should serve him: his dominion is an everlasting dominion, which shall not pass away, and his kingdom that which shall not be destroyed. (Daniel 7:8-14)

God demonstrates His righteousness by prolonging the life of the rebellious who remain alive after the Day of the Lord's Wrath. Scripture portrays the coming thousand-year reign of Christ with believers as a strict though compassionate dominion.

Together, Christ and the church comprised of Old and New Testament believers' rule over the unregenerate nations of the world as demarcated by the stern Kingdom of David and peaceable Kingdom of Solomon.

Thou shalt break them with a rod of iron; thou shalt dash them in pieces like a potter's vessel. Be wise now therefore, O ye kings: be instructed, ye judges of the earth. Serve the Lord with fear, and rejoice with trembling. Kiss the Son, lest he be angry, and ye perish from the way, when his wrath is kindled but a little. Blessed are all they that put their trust in him. (Palms 2:9-12)

Let the saints be joyful in glory: let them sing aloud upon their beds. Let the high praises of God be in their mouth, and a two-edged sword in their hand; to execute vengeance upon the heathen, and punishments upon the people; to bind their kings with chains, and their nobles with fetters of iron; to execute upon them the judgment written: this honour have all his saints. Praise ye the Lord. (Psalms 149:5-9)

After the millennium, these nations, though living under benevolent rule and the reign of Christ and the saints for a thousand years, when given the alternative of submitting to Christ or Satan, choose the latter. But then, just as David slew the Amalekite (2 Samuel 1:15,16) and Solomon destroyed Adonijah (1 Kings 2:24,25), Satan must then be finally vanquished, and cast into the lake of fire, which is the final death closing the millennium.

Yet all, righteous and rebellious, must ultimately stand before the Great White Throne judgment of God in heaven (Revelation 20:11-15). There the books will be opened and the saints justified by Christ's sacrifice while the words and works of the ungodly bring their judgment.

Cosmic Signs

Returning to the opening of the seals, a parallel can be drawn between the description of cosmic signs in the Olivet Discourse of Matthew 24 and the opening of the sixth seal in Revelation 6. Jesus stated regarding the period of Great Tribulation commencing after the Abomination of Desolation, that no flesh would be saved except those days should be shortened (Matthew 24:22). Referring to this abbreviation of the Great Tribulation, Jesus declared,

> Immediately after the tribulation of those days shall the sun be darkened, and the moon shall not give her light, and the stars shall fall from heaven, and the powers of the heavens shall be shaken: And then shall appear the sign of the Son of man in heaven: and then shall all the tribes of the earth mourn, and they shall see the Son of man coming in the clouds of heaven with power and great glory. And he shall send his angels with a great sound of a trumpet, and they shall gather together his elect from the four winds, from one end of heaven to the other. (Matthew 24:29,31)

The parallel passage at the opening of the sixth seal paints a similar picture of a cosmic disturbance:

> And I beheld when he had opened the sixth seal, and, lo, there was a great earthquake; and the sun became black as sackcloth of hair, and the moon became as blood; and the stars of heaven fell unto the earth, even as a fig tree casteth her untimely figs, when she is shaken of a mighty wind. And the heaven departed as a scroll when it is rolled together; and every mountain and island were moved out of their places. (Revelation 6:12-14)

On these two passages the primary theme of this book pivots. The blessed hope (Titus 2:11-14) refers to a joyous expectation of the Lord Jesus Christ appearing. Christ will indeed come again as He departed. First, He will appear in the heavens to raise the righteous dead and catch away believers from the present fallen world. Then the wrath of God will fall during a period called the Day of the Lord.

This is a great mystery. Yet there are aspects of His coming known because they are revealed or discernable. Scripture provides a revealed framework of cosmic events surrounding Christ's return. These are not isolated occurrences, but rather part of a sequence of signs emanating from the initiation of a singular action. Neither are the events natural in the sense of resulting from a physical cause.

Isaiah 64 opens with a plea that ties into the description of heaven and earth as developed in Chapter 11 of this book with the creation account:

> Oh that thou wouldest rend the heavens, that thou wouldest come down, that the mountains might flow down at thy presence. (Isaiah 64:1)

In the same way, David prophesied in Psalm 18:

> Then the earth shook and trembled; the foundations also of the hills moved and were shaken, because he was wroth. There went up a smoke out of his nostrils, and fire out of his mouth devoured: coals were kindled by it. He bowed the heavens also, and came down: and darkness was under his feet. …Then the channels of waters were seen, and the foundations of the worlds were discovered at thy rebuke, O Lord, at the blast of the breath of thy nostrils. He sent from above, he took me, he drew me out of many waters. He delivered me from my strong enemy, and from them which hated me: for they were too strong for me. (Psalm 18:7-9, 15-17)

And also in Psalm 144:

> Bow thy heavens, O Lord, and come down: touch the mountains, and they shall smoke. (Psalm 144:5)

This opens the door to the Day of the Lord. In addition to the earlier reference in Matthew 24, cosmic events are prophesied elsewhere in Joel, Zephaniah, 1 and 2 Thessalonians, and Revelation.

Various descriptions including signs in the sun, moon, stars, and earth seem somewhat disconnected on the surface. Yet upon closer examination, they all emanate from the heavens rending open to introduce the return of Christ.

Going back to the earlier discussion, the created universe can be described as a spacetime fabric, with mass and energy embedded and interacting as represented by Einstein's relativistic field tensor. This fabric corresponds to the description provided in Matthew 24. Isaiah and Jeremiah denote the heavens rending, stretched, and rolled up.

Following is a sequence of observable happenings resulting from the rending open of the heavens.

The Earth Trembles

In recent years a scientific endeavor called LIGO-Virgo collaboration has been developed to detect major gravitational disturbances in the universe. LIGO (US based pair of Laser Inferometer Gravitational Wave Observatories) and Virgo (complementary European observatory) sense gravity waves originating from beyond the earth.

Anyone who has experienced an earthquake knows the feeling of a rolling or shaking wave passing under foot. Quake detectors located in various parts of the earth determine the location, timing, and strength of seismic occurrences around the world.

Identifying subtle gravitational shifts emanating from cosmic disturbances requires a unique sensing methodology referred to as Laser interferometry with lasers targeted at sensors miles apart for sensitivity.

By sensing these disturbances at two widely separated earth-based locations, LIGO-Virgo can determine whether the event originates from beyond the earth and from what location. A celestial disturbance can then be identified with an observed astronomical event.

References to the beginning of the Day of the Lord including Joel

(3:15-16), Isaiah (13:10-13), Matthew (24:29), Mark (13:25), Luke (21:26), Revelation (6:12), and Hebrews (12:26) describe a great shaking of heaven and earth. Many of these Scriptures denote tearing or rolling up of the heavens like a scroll following a terrestrial great earthquake. The implication of rending the heavens precipitates a worldwide shaking on earth. As the Lord said:

> …Yet once, it is a little while, and I will shake the heavens, and the earth, and the sea, and the dry land; and I will shake all nations, and the desire of all nations shall come: and I will fill this house with glory, saith the Lord of hosts. (Haggai 2:6,7)

At the first indication of a major earthquake, virtually everyone makes their way outside because of the risk of a building collapse. In the opening before the Day of the Lord, a worldwide major quake will have the result of forcing populations to immediately go outside. Visible to all the earth there will be a rapidly unfolding succession of unprecedented cosmic events.

Before looking at these events individually, consider how this will affect everyone in the world simultaneously. While the rending of the heavens and subsequent quaking will touch everyone around the world immediately, can the same be said for visually observable cosmic events?

First, most of our understanding of the heavens is based on observable measurement rather than physical sensing. This includes visible electromagnetic radiation (light spectrum from infrared to ultra-violet) as well as radio waves (longer wavelength) and x-rays or gamma rays (short wavelength). Until recently the heavens were perceived only by visible radiation. Understanding these observables utilizes mathematics (symbolic logic) and theoretical modeling to explain a phenomenon or to predict a future event under specified conditions.

Only within the past sixty years, space probes have travelled to the moon, Venus, Mars, asteroids, as well as interstellar space to make measurements that do not solely rely on electromagnetic waves.

Results are still returned in the form of electromagnetic signals. In the case of lunar landings, some physical samples have been returned for inspection. When the Day of the Lord unfolds, will the world

Chapter Seventeen - Second Coming

depend on television, video camera, or cell phone transmissions to observe these events?

To answer that question, consider what occurred in the battle supporting the Gibeonites against the Amorite nations (Joshua 10:12-13). Joshua spoke to the Lord and commanded the sun and moon to stand still. Verse 14 says, "And there was no day like that before it or after it, that the Lord hearkened unto the voice of a man."

The names Jesus, Isaiah, Joshua, and Hosea all derive as an abbreviation from the Hebrew name Jehoshua (meaning Jehovah saved). Joshua declared, "...Sun, stand thou still upon Gibeon; and thou, Moon, in the valley of Ajalon." (Joshua 10:13)

God Himself responded to the assertion of Joshua as a type of Christ to foreshadow events at the opening of the sixth seal. Similarly, in Hezekiah's interaction with Isaiah, he was given the option for a sign of the sun's shadow lowering or returning ten degrees on the sun dial of Ahab (Isaiah 38:8).

These biblical events are occasionally referred to falsely as a "missing day" which NASA supposedly confirmed. No such confirmation ever happened. The "Missing Day Verification" story, though often repeated, is a complete fabrication.

Many biblical narratives describe miraculous events. As written, the battle in Joshua continued while the sun and moon appeared to remain fixed in the sky for about a day. What was happening in the heavens?

For anyone with a knowledge of orbital mechanics, that is problematic. While it can be argued, God is not limited by physical processes, and can do anything He wants, we know stopping the movement of the sun and moon would involve quite a complex set of maneuvers.

Technically the earth would need to decelerate and then reaccelerate in this case. With the earth's circumference of roughly 25,000 miles rotating at one revolution each 24 hours, that is a surface velocity of just over 1000 miles per hour. The resulting tidal shift of a rotational stop/start would be traumatic—let alone all the other implications of such an event. Something else must be going on.

In the story of Hezekiah and Isaiah, the offered sign was framed

191

as the shadow returning, rather than the earth turning back or sun changing directions. If we consider spacetime as a fabric, the sun and moon can appear to stand still simply by a distortion of the spacetime fabric.

That is no less a miraculous maneuver, but makes much more consistent sense in the description context of what happens in the books of Joshua and Isaiah and at the end of the age. In both cases, spacetime distorts by stretching or rending of the heavens.

As the heavens are rolled up as a scroll, light and electromagnetic radiation in transit will be distorted and appear as if the sun, moon and stars are moving across the heavens. As described in Matthew 24:29, "...the stars shall fall from heaven...."

The movement of light from the stars will appear as though they are falling across the skies. Coincidently, recent observation of distant galaxies with the James Webb Space Telescope (JWST) illustrates how light bends around large masses, in a process termed gravitational lensing, a distinct but similar occurrence.

Consider a further characteristic of mass, spacetime, and the movement of light or radiation based on Einstein's relativistic field equations. Objects are characterized by mass, and an associated contribution to spacetime.

The influence of all other mass in the universe also acts on every object by contributing to the vacuum energy density of space characterized by spacetime fabric or tension on the continuum.

Even without the influence of other mass, each object initiates its own spacetime contribution. This characteristic results in relativistic differences of spacetime measure when an object moves at or near the speed of light. At one point in his development of the tensor field equation, Einstein added a Cosmological Constant to create a relationship which predicted a non-expanding universe based on the equation. It was not until Edwin Hubble demonstrated the universe was expanding based on the redshift of light, that Einstein declared his use of the constant a mistake.

If the bound of the heavens and fabric of space is rent, and rolled up, that is a non-local influence. Local spacetime created by sun, earth, and moon still provide dominant influence on localized spatial fabric.

But light in transition from stars and sun particularly are shifted. And as the spatial fabric of the heavens quickly rolls back, light from the stars, sun, and moon appear to move across the sky with everyone across the world witness to the result. Look at more implications of the heavens rolling away.

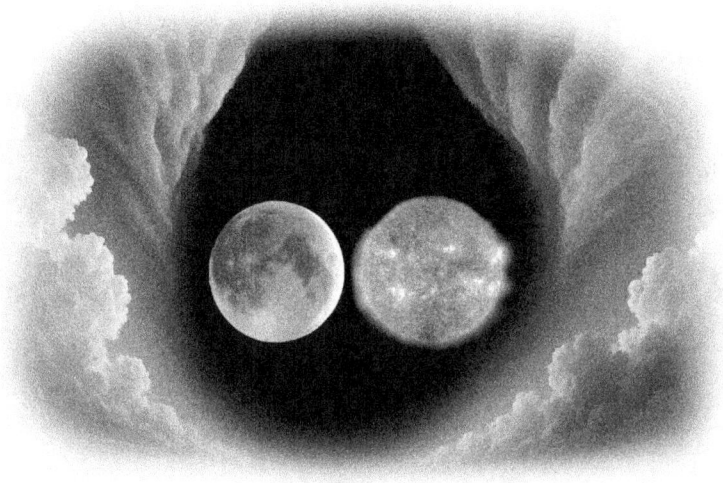

The Heavens Roll Away as a Scroll

Arch of the Heavens

In the creation account there are several descriptions about the nature of the heavens. At the beginning, heaven and earth are described as without form, void, and dark while the Spirit of God moved on the face of the waters. With the Word of God came the creation of light and the separation of light and darkness defining the first day and night. Again, the Word of God spoke into existence a firmament within waters on the second day, dividing the waters above and below.

And God said, Let there be a firmament in the midst of the waters, and let it divide the waters from the waters. And God made the firmament, and divided the waters which were under the firmament from the waters which were above the firmament: and it was so. And God called the firmament Heaven. And the evening and the morning were the second day. (Genesis 1:6-8)

What is meant by waters? The Hebrew term *waters* or *mayim* (meaning water or fluid) appears straightforward. The Hebrew for *heaven* or *shamayim* means "sky" or "lofty waters." On the fourth day God said, "...Let there be lights in the firmament of the heaven to divide the day from the night; and let them be for signs, and for seasons, and for days, and years." (Genesis 1:14)

The sun, moon, and stars were set (from the Hebrew term *nathan* meaning given) to establish light on the earth and to create the division of day and night. Light was not created in process, so the firmament would have stretched out on the fourth day as the stars formed.

The expression, "...a firmament in the midst of the waters, and let it divide..." paints an image of the extent of the heavens bounded at its' limit and then stretched out. This concept has a different connotation from that of a supposed vapor canopy layer above the sky proposed by Isaac Vail in the early 1900s, later adopted by some creationists to justify certain hypothetical preflood conditions.

When God stretched out the heavens, the composition bounding the perimeter of the sphere was not described beyond waters. The use of the Hebrew word *shamayim* or "lofty waters" does not qualify an exact nature or understanding of the waters constituting the outer limit of the firmament.

With the development of sensitive, low frequency radio telescopes, a pervasive noise could be detected across the skies. In 1965, while calibrating the Holmdel Horn Antenna, radio astronomers Arno Penzias and Robert Wilson identified the source as microwave background glow.

It required several decades for radiometric sensors to be able to fully map cosmic background radiation. By 1992, measurements from the NASA Cosmic Background Explorer (COBE) satellite were published as the Cosmic Microwave Background (CMB).

Refinement of the CMB continued with the Atacama telescope, Planck Surveyor satellite, and South Pole telescope. The latest results give a color temperature range of the CMB as 2.7260+/-0.0013 Kelvin, representing very close to absolute zero.

Several forms of matter can potentially be a candidate for makeup of a perimeter of matter bounding the heavens. Because temperature based on CMB radiation approaches absolute zero, options are limited.

A Bose-Einstein Condensate (BEC), in principle, must be made of bosons, and formed when a system of particles is cooled even closer to absolute zero, ~38 pK (10^{-12} K). That would seem to exclude water molecules forming a BEC, as water cannot be bosons, in their typical form. Alternatively, water ice assumes unique structural properties when cooled near absolute zero.

The best determination of this perimeter would be something like a boundary of water, or form of matter providing considerable mass as a boundary of the heavens. Regardless of the boundary makeup, the specifics are secondary to the central point that it serves as the mechanism of mass with means to stretch out the cosmos.

Because Scripture does not provide a clarification of the makeup, the term "waters" for this bound suffices. Other Scriptures support the Genesis 1 account of waters bounding the cosmos. Elihu reproved Job with the question:

Hast thou with him spread out the sky, which is strong, and as a molten looking glass? (Job 37:18)

In the Psalms, the waters above the heavens are referenced:

Praise him, ye heavens of heavens, and ye waters that be above the heavens. (Psalm 148:4)

Also, in Revelation:

And before the throne there was a sea of glass like unto crystal... (Revelation 4:6)

And I saw as it were a sea of glass mingled with fire... (Revelation 15:2)

And in 2 Peter:

And saying, Where is the promise of his coming? For since the fathers fell asleep, all things continue as they were from the beginning of the creation. For this they willingly are ignorant of, that by the word of God the heavens were of old, and the earth standing out of the water and in the water: Whereby the world that then was, being overflowed with water, perished: But the heavens and the earth which are now, by the same word are kept in store, reserved unto fire against the day of judgment and perdition of ungodly men. (2 Peter 3:4-7)

Beyond this bound there would be nothing of the physical world; no space, no time, no mass, no light, at least not in a natural sense. But these "waters" create gravitational tension on the interior heavens. To express this in a more technically correct manner, mass causes a distortion of spacetime, which in turn results in mass moving in a way that appears to give rise to gravitational force.

That raises the question about the effect of gravitation within a spherical shell; the classical shell theorem. A hollow sphere does create a gravitational potential exterior to the shell but not within. The net potential inside can be shown to be zero. But notice there are two unique characteristics to this shell bounding the heavens.

First, space and time are constrained within the shell. From the relativistic field equation, even a classical shell in space creates an influence on the spacetime inside as a result of its' mass. This is not a gravitational potential, but a distortion of spacetime similar to the tension on the fabric stretched over a drumhead.

Second, this sphere is not merely a classical shell in space but a bound of space and time. How does this influence the light radiated from bodies within the sphere, particularly near the perimeter? Modeling of the spacetime continuum with the relativistic field equation simulation is not a simple exercise and well beyond the scope of this writing.

Finally, and most importantly, increasing redshift with distance infers stretching a representative of spacetime influenced by mass bounding the observable universe. A key visual observation is the redshift of light which increases with distance.

Cosmologists currently hypothesize this results from dark matter and dark energy distributed within the universe. The intent is to explain observed acceleration toward the peripheral bound or edge of the cosmos. The creation account presents these bounding waters as the matter continuing to stretch out the fabric of spacetime and resulting redshift of light.

Cosmologists do not have clear and convincing evidence of the existence of any other form of dark matter or dark energy; only that something provides a mechanism like Einstein's cosmological constant to explain observed behavior of the universe. The Lord took Job to task when he asserted that God owed him an explanation, saying, "Where wast thou when I laid the foundations of the earth?" (Job 38:4).

Scripture presents a compelling picture of the unveiling of creation, the stretching out of the heavens, and an explanation for the observed universe. In addition, this serves as a means to illustrate how the events of the Lord's return unfold consistent with the biblical account.

Mass constituting the outer bound to the cosmos explains the observed stretching rather than the need to postulate dark matter. Also, the boundary emits radiation at the average color temperature characterizing the CMB, close to absolute zero.

Boundary of the Heavens

An outer bound to the heavens formed by "waters" or the Hebrew word "*mayim*" can be envisioned as a spherical limit to mass, space, light and time. What form would it take? What would its mass be relative to the visible universe? The questions can be benchmarked by a simple calculation. In order to calculate a basic estimate of boundary thickness, make the following assumptions,

1. The current estimated radius of the visible universe is 46.5 billion light years.
2. Composition of the boundary is water.
3. Total mass of the boundary constitutes estimated dark matter (15% visible / 85% dark).
4. Estimated mass of the visible universe is 1.5×10^{56} gm and dark matter 9.0×10^{56} gm.

Surface area of the sphere would be:

$$A_s = 4\pi R^2$$
$$= 4 \times 3.1416 \times (46.5 \times 10^9)^2 \text{ LY}^2$$
$$= 2.72 \times 10^{22} \text{ LY}^2 \text{ (LY = Light Year)}$$
$$= 2.43 \times 10^{58} \text{ cm}^2$$

Since water density $D_w = 1.0$ gm/cm^3, the thickness would be:

$$T = V/A_s = M/(D_w A_s)$$
$$= 9.0 \times 10^{56} \text{ gm} / (1.0 \text{ gm/cm}^3 \times 2.43 \times 10^{58} \text{ cm}^2)$$
$$= 0.037 \text{ cm (or 0.015 in)}$$

That might seem extraordinarily thin, but note the sphere bounds a complete universe, where net volumetric density is comparatively low, similar to a bubble sustained by surface tension. In this case space and time are confined within.

Slough of Despond

To gain insight into the structure of the universe bounded at the perimeter, it would be helpful to represent spacetime fabric with Einstein's field equation tensor and seek a solution. That can only be done for very simplified conditions. We are again challenged by Mount Difficulty of John Bunyan's *Pilgrim's Progress*.

Instead, it may be prudent to do something transformational and take another path, temporarily and figuratively descending through Bunyan's Slough (pronounced slew or slow) of Despond. In Florida transiting a slough or swamp as a shortcut also gets messy.

Theorizing a universe spacetime, sparsely populated with matter and energy and bounded at the limit, can be challenging, let alone to model. What is the center? What constitutes the boundary? What are the constraints? In particular, with no space time beyond the bounding matter, what boundary conditions apply?

Instead try something transformational to simplify the problem. In order to perform a transformation of spacetime fabric into a complex number space requires some mathematical gymnastics that bog down in the Slough of Despond, so take the bridge instead and enjoy the view.

As part of graduation requirements for a Master's Degree at the University of Cincinnati, a published thesis project was completed, "Study of Incompressible Viscous Flow in a Pipe-Orifice Using Unsteady Navier-Stokes Equations and Fully Implicit Numerical Method," 1984. Part of the challenge of doing the analysis required the transformation of complicated flow geometry into a simplified axi-symmetric grid in order to perform difficult computations with time dependent fluid flow equations.

That is a mouthful, but at least it makes computation easier on the processor, mapping to a complex variable geometry. Once a computation converged, the solution was mapped back to real spacetime through the reverse transformation.

Reading the thesis again after forty years, through the mathematical and programming gymnastics, it becomes apparent many of the required neurons and synapses have been usurped to remember names, birthdays, and passwords.

Key to the successful thesis strategy involved the development of the series of clever spacetime transformations of orifice geometry for computations. The thesis acknowledgements began:

"I would first like to offer thanks to God, through whose Son Jesus Christ my life has been transformed, for the grace to carry this project to completion...."

Upon reading the thesis acknowledgements, my thesis committee chairman exclaimed, "What do you mean by transformed?" Glad you asked.

On this bridge we avoid messy mathematics and view the conceptual result. A mathematical transformation of inversion results in all of the boundary matter (cause of observed redshift stretching) becoming a spherical center of spacetime.

Rather than bounded at the perimeter, spacetime now becomes

unbounded, extending to infinity (mathematically). All other matter-energy in the form of celestial bodies and interstellar dust occupies spacetime above the central matter. Spacetime and matter-energy are stretched toward the central mass and characteristic of matter moving toward a gravitational well, releasing radiation of observed CMB.

Realistically, if the transformed space was for computation of the Einstein field equation tensor, such a concentration of mass at the center would seem to create a messy black hole singularity. But the center of this transformation would represent only the outer bound of spacetime and distribution of a solution, not physical properties of the original problem.

The outcome means characteristics of the Genesis 1 account of creation with spacetime bounded by mass-energy at the perimeter as well as contemporary observation of stretched light redshift can be satisfied conceptually.

After fifty years reading and reviewing popular as well as technical articles on cosmology, rarely have any taken the approach of reconciling the Genesis account, other than to treat it dismissively.

A survey of cosmology publication author bios reveals the vast majority are culturally of Jewish heritage, yet most profess secular or atheistic viewpoints as during my teen years. It would be enlightening to see an analytical study of the configuration with a fair, impartial investigation of the implications.

Unfortunately, non-secular worldviews in literature are celebrated like a strikebreaker in Union Hall, similar to what was experienced after assembling the CFM56-2 gearboxes.

Sun as Sackcloth

Returning to the simulation of spacetime as a stretched fabric with spherical metal weights on the surface, consider what happens when tension is increased or decreased on the perimeter. This would simulate the influence of changing attraction from distant mass. As remote influence decreases with the rending of the heavens, the local spacetime tension decreases, resulting in the stretching of wavelengths of light in transit.

How does this affect observations? Anticipating the first objection, does that mean the impact of rending a bound of "waters" at the perimeter would require thirteen billion years for the influence of spacetime and light to reach earth?

On the premise of the argument, God created the heavens on the second day (Genesis 1:6-8), stretching out with a bound of "water" above and below on the fourth day. As a result, the "firmament" between is stretched to define spacetime and distribution of mass-energy as observed.

Similarly, God's purpose at the opening of the sixth seal represents a work to initiate the removal of the faithful and dead in Christ as well as judgment during the Day of the Lord's Wrath. It does not constitute a deterministic outcome of events already started at the bounding surface, but of an immediate response to God's fiat action, rolling out the cosmic scene to everyone across earth in rapid succession.

Integral to the event, with sunlight shifted to infrared the sky becomes black, a darkened sackcloth sun, blood red moon, traversing stars, the scene rolls quickly around the earth. How does this materialize from the stretching of the heavens?

First consider that each celestial object has a different spectral signature. Light is emitted based on temperature and composition or it is reflected subject to light source, composition, and reflectivity.

There are other factors such as object size, redshift due to motion, or the effect of clouds, for example, on Jupiter and Saturn. In addition, light experiences absorption lines resulting from elements on the surface or in the path to earth.

The nature of these absorption lines gives information about the surface composition or material in interstellar space. Our sun has its own unique spectral signature measured in space above earth as well as on the earth's surface. As with all stars, the sun's spectrum ranges in wavelength and intensity from infrared to ultraviolet, with the preponderance of solar energy in the visible to ultraviolet.

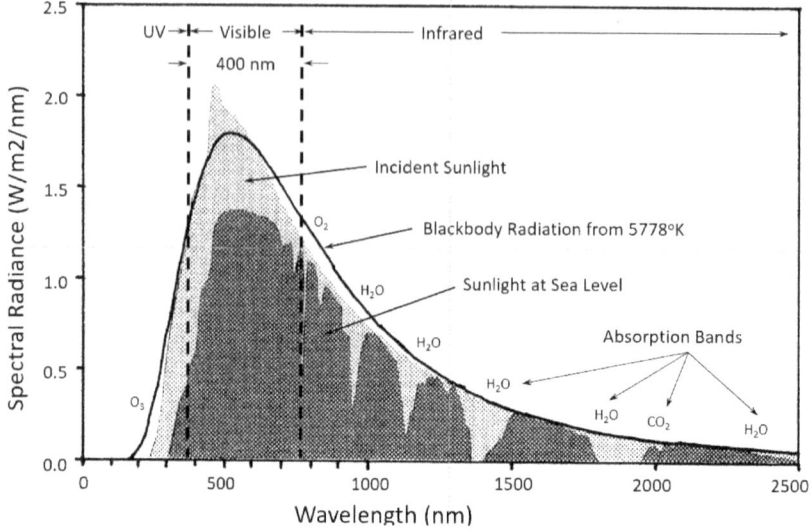

Chart of Solar Radiation Spectrum vs. Wavelength

This occurs as a function of temperature and makeup of the sun's surface and corona where radiation originates. At the surface of earth, specific wavelengths of light exhibit lower intensity resulting from energy absorption and scattering.

Ultraviolet light dispersion due to oxygen (O_2) and ozone (O_3) represents the most obvious example of scattering, giving rise to blue skies in daylight. Outside the atmosphere, solar radiation contains damaging levels of UV radiation, and lacking scattering, the sky appears totally black rather than blue.

Moving to lower altitudes skies transition from black, to indigo, to blue as more oxygen molecules scatter UV light. Atmospheric conditions at the Dead Sea (~400 m below sea level) create unique circumstances of low humidity and low UV light favorable for skin phototherapy.

When frequencies of light shift due to the stretching of spacetime, unique changes take place as the preponderance of ultraviolet transitions toward infrared.

Loss of blue wavelengths means the sky in daylight becomes completely black, typical of what is experienced outside the atmosphere from the stratosphere and beyond. Also, as sunlight becomes mostly infrared, the sun takes on a dark red hue in a black sky. So, a reddish-black sun transitions across a night sky as the heavens roll quickly away around the world. That reflects the biblical description in both the Old Testament and New Testament accounts of characteristic signs preceding the Day of the Lord. Next, what happens to the appearance of the moon?

Blood Red Moon

Recent prophetic teachers suggest a series of so-called blood moons prior to the Lord's return. These notions do not match biblical narratives before the Day of the Lord. Descriptions of a blood red moon in Isaiah, Joel, Matthew, and Revelation associated with the Day of the Lord represent something entirely unique. First, consider what happens in the context of events at the Lord's return in the heavens just prior to the Day of the Lord's wrath.

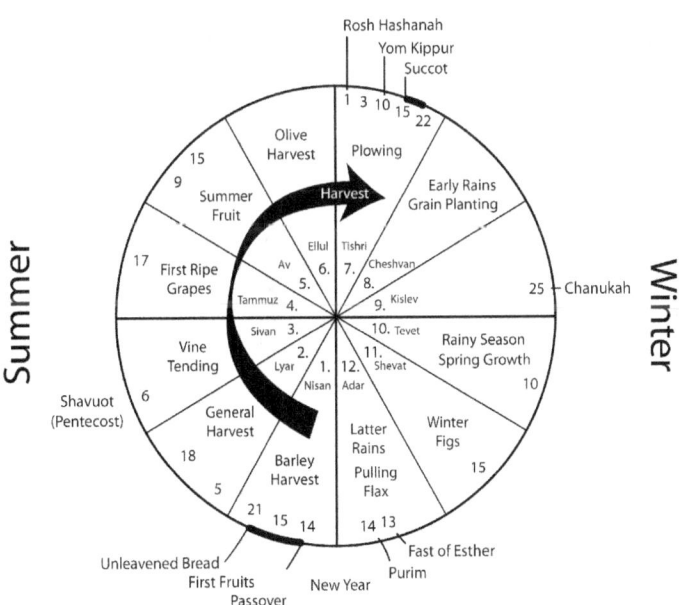

Circular Chart of Jewish Yearly Calendar

Feasts of Passover Season

Of seven "Festivals of the Lord" in Leviticus 23, four have been completed or fulfilled. Each of the seven is a symbolic representation based on the past as a precursor of the future. Passover Season, instituted symbolically in the exodus from Egypt had fulfillment with Christ's sacrifice on the cross. At the celebration of Passover, three festivals are satisfied. Passover, First Fruits, and Unleavened Bread were symbolized in Egypt and completed in Christ during the season of Passover at Christ's Passion and resurrection.

Festival of Weeks

As noted in Exodus 19, on the third month and third day, the people prepared themselves to receive the law by the hand of Moses. Coming of the law written in stone fifty days after Passover was symbolic of the work of the Holy Spirit when, on the Day of Pentecost, the law was revealed to the heart. Hence the fourth festival also had fulfillment symbolically and literally at the Jewish Festival of Weeks.

Feast of Trumpets

For this reason, fulfillment of Christ's return in the heavens would be expected at the Festival of Trumpets. This occurs the first day of the seventh month of the Jewish religious calendar (first day of the first month of the civil year). How does this reconcile with what Jesus declared?

But of that day and hour knoweth no man, no not the angels of heaven, but my Father only. (Matthew 24:36)

Consider that Rosh Hashanah (Trumpets, or literally Head of the Year), though a set day each year based on the current Jewish calendar (163 days after Passover), was not traditionally identified until the High Priest certified the first new moon of the civil year.

Observing the absence of the moon during a new moon becomes difficult near the sun. Apart from new moon occurrences associated with a partial or full solar eclipse, a new moon would appear as a sliver of crescent adjacent to the sun.

The moon's orbit of earth lies in a plane tilted about five degrees from the ecliptic, which is Earth's plane of rotation about the sun. Therefore, the moon often appears out of plane from the sun when observing a new moon crescent. Even under ideal conditions just after sunset, observing the new crescent moon remains quite difficult to see.

Therefore, no one "knew the hour or day" until signaled by the High Priest based on an observable sign seen by two witnesses, though "times and seasons" (I Thessalonians 5:1) were known then, as they are observed today.

Once declared near sunset, bonfires on hilltops and the blowing of trumpets signaled Rosh Hashanah to the nation of ancient Israel. A seeming inconsistency of the new moon at Rosh Hashanah and these cosmic events would be the description of a blackened sun and blood red moon with rending and rolling up of the cosmic spacetime fabric.

Normally a new moon cannot be seen with ease against the backdrop of the sun, unless there is reflected earthlight to illuminate the moon. This difficulty is compounded by the brilliance of the sun, but not if the sun's light is darkened by the frequency or wavelength shift to infrared.

Now the challenge is still one of reduced sunlight intensity or luminosity because the light spectrum has shifted from ultraviolet to infrared. Though the moon and sun are roughly in conjunction, the dilemma is not resolved by the conditions of a solar eclipse.

It would seem difficult for the moon to appear so conspicuous without further explanation. Is it possible the moon has greater reflectance to infrared wavelengths than visible or ultraviolet? More to come on this shortly.

The first clear and consistent appearance of believers caught away occurs after sealing of the one hundred forty four thousand who are protected against wrath to come. John declared the assembly he saw in heaven:

After this I beheld, and, lo, a great multitude, which no man could number, of all nations, and kindreds, and people, and tongues, stood before the throne, and before the Lamb, clothed with white robes, and palms in their hands; and cried with a loud voice, saying, Salvation to our God which sitteth upon the throne, and unto the Lamb. (Revelation 7:9-10)

All classes or categories of people are included with two notable exceptions. Absent are mention of "all beliefs or faiths," nor provision for "all moral distinctives," i.e. good and evil. "Beliefs" and "actions" are two qualities of character that separate sheep and goats; those who inherit eternal life verses those who inherit death based on matters of choice, or Hebrew *shaal* meaning, "That's what you demanded."

Feast of Yom Kippur

The Day of Atonement (Yom Kippur) follows Rosh Hashanah on the tenth day of the seventh month, representing the highest holy day of the Jewish calendar. A unique feature of Yom Kippur relates to the day of general repentance, sacrifice, and atonement of the Jewish people, both symbolically as yearly celebration and a final day of the seven-year period when God brings the remnant of one hundred forty-four thousand face to face and the knowledge of Jesus Christ as Jewish Messiah.

Basis for the Jewish festival calendar rests upon a "luni-solar" system. A lunar month based on astronomical observation is approximately 29-1/2 days, yielding a 354-day lunar year. In order to match the solar year of roughly 365 days, a leap month Adar is added seven times every nineteen years. In the book of Daniel, as well as elsewhere, an even 30-day month is used for a 360-day prophetic year.

In Daniel 9, God revealed through the archangel Gabriel, a period of seventy sevens (490 years) or seventy *shabua* (meaning seven) are determined upon Jerusalem (Daniel 9:25-27). Of these, seven plus sixty-two weeks of years (or 69 x 7 = 483 years) from the command to restore Jerusalem until the Messiah the Prince. Here a year entails the 360-day Jewish prophetic calendar year.

Prophetically, the Jewish calendar period matches to the day from Artaxerxes decree to Nehemiah for rebuilding Jerusalem to the triumphal entry of Jesus on Palm Sunday, 33 CE (Christian Era, Zechariah 9:9). That day represented completion of the promise for the revelation of the Messiah.

There remains a final week or *shabua* of years upon God's people the Jews. Regrettably, the final *shabua* frequently has been referred to as "The Tribulation." For thousands of years God's people, both Jewish and Christian, have experienced tribulation. That's nothing new.

At any given historical time and world region, believers suffer persecution representing loss of livelihood, family, finances, homes, places of worship, health, and life itself. Though statistics vary depending on the definition of persecution, a large percentage of believers today face what can be called true tribulation, though limited in space and time.

Recent reports of Christian beheadings represent a foreshadowing of things to come at the opening of the fifth seal in Revelation 6:9-11. These three verses encapsulate the period Jesus referred to as a time of "Great Tribulation" (Matthew 24:21,22), when both Christians and Jews throughout the world will suffer affliction, torture, and death of a greater scale than the Maccabean or Nazi Holocausts.

Great tribulation commences with the introduction of the Abomination of Desolation defiling sacrificial service in a temple or tabernacle, and interrupted at the sixth seal with the appearance of Jesus Christ in the heavens. At that point action shifts dramatically from the Antichrist inspired great tribulation to *YHWH* implemented great wrath.

This restricted period of months or "little season" (Revelation 6:9-11) encapsulated in these three verses starting the mid-point of the *shabua* represents the most severe and universal persecution of Christians and Jews in history. As Jesus noted:

> And except those days should be shortened, there should no flesh be saved: but for the elect's sake those days shall be shortened. (Matthew 24:22)

The Archangel Heylel, failing to elevate himself above God Almighty and cast down, at this point through his minion the Antichrist, stands on the precipice of achieving his ultimate goal, the destruction of Jews and Christians to thwart the kingdom of God and Christ.

Yet for the unregenerate remaining after the catching away at the sixth seal, true wrath and destruction reflect the worst of judgments to follow during seven trumpet judgments and seven vials.

Many teachers adopt a position insisting Christians depart before "tribulation," or the seven years, reasoning the alternative would discourage the saints. For souls under the altar, slain for the Word of God and their testimony, Jesus did not offer concern their sacrifice might discourage fellow servants and brethren facing the same death.

Masking the problem by inventing a label "tribulation saints," saved after the catching away, imposes an artificial construct unsupported in Scripture. When confronted with danger, face it head on and make the problem part of the solution. Today believers around the world face that prospect daily. If discouraging the saints represented a legitimate doctrinal backstop, why not stop talking about sin and ignore hell?

The intent here rests on clarifying a consistent sequence of events as described in Scripture. It does not hinge on a breakdown and comparison against the spectrum of teaching alternatives. Scripture regarding Christ's return stands on its own merit as to what God means when He speaks, and by the fulfillment of events themselves.

After the signing of an agreement referred to as a "covenant with death" and the unveiling of the Antichrist, the reality will be painfully obvious. Hundreds of millions of Christians around the world are already experiencing what can be referred to as severe tribulation, or the "beginning of sorrows." Yet all true believers are to be caught up before the impending wrath of the Day of the Lord, constituting events after the opening of the seventh seal (Revelation 8:1).

A final week to come can be identified with the Nation of Israel submitting to a supposed peace agreement or "covenant with death" (Isaiah 28:14-19) for the guarantee of their protection.

In Daniel 9:26, speaking of a "prince who is to come," he (the Antichrist) confirms a covenant with many for a *shabua* or seven. That may constitute ratification of an existing agreement. Scripture references the period of seven as two halves of 3-1/2 years (42 months, 1260 days) based on a 360-day Jewish prophetic lunar calendar.

In the middle of the seven, the Antichrist breaks the covenant of peace, just as did Antiochus IV Epiphanes with the Maccabees in the inter-testamental period. That description of the Abomination of Desolation foreshadows the future referred to by Jesus during the Olivet Discourse (Matthew 24, Mark 13, Luke 21).

Abraham's covenant with Abimelech (Genesis 21:22-33) at Beersheba (well of the oath, oath of seven) provides foreshadowing of the current Abraham Accords, seeking to reconcile nations, peoples, and creeds of surrounding nations to God's covenant people. Drawn by a promise of security, Israel will be led into a compromising trap as the Antichrist rises to a place of influence.

At some point in the process, the signing of such a covenant agreement by the Nation of Israel marks the commencement week or *shabua* of 2 x 1260 days (2520 days) until the ultimate fulfillment of the Day of Atonement, when one hundred fort-four thousand Jewish remnant, hidden in Sela or Petra, are visited by a glorious appearing of Jesus their Messiah (Revelation 14:1-5) and come to a personal knowledge of Him (Zechariah 12:9-14).

At this time Christ, and the remnant of Jewish believers will follow Him to confront the Antichrist and the unregenerate nations. Simultaneously, in the final outworking of God's wrath, seven angels go forth to pour out seven vials or bowls of the wrath of God Almighty (Revelation 16:1-21).

Each fall for over thirty years, I have calculated the day 2520 days before Yom Kippur seven years off to monitor whether a peace covenant signing takes place. There is coming a year when a covenant will be signed on that day signaling the commencement of those final seven years during which the seals will be opened. For the year 2025, that day 2520 days before Yom Kippur 2032 will be October 20-21, 2025.

After the date of that covenant, at the midpoint of 1260 days, a desecration of the place of sacrifice will take place, ushering in the Abomination of Desolation and period of great tribulation. These events, revealed to Daniel and John and sealed in mystery, underpin the plan of God for Christ's return. Whether 2025 or the following year, once the covenant of seven is signed for Israel's protection, the clock will begin a countdown toward the Abomination 3-1/2 years later and Rosh Hashanah about five years later.

Times and seasons are known, though closure of the final act remains shrouded in the purpose of God Almighty. As events begin to unfold, they execute according to strategic steps of a plan rather than disconnected chaos. Jesus challenged all therefore to;

And what I say unto you I say unto all, Watch. (Mark 13:37).

Lunar Reflectivity

Upon posing a question to fellow scientists about lunar reflectivity, a number of technical reference papers were recommended. After some research, a study of lunar light yielded a publication and data showing the variation of spectral reflectance as a function of wavelength for a range of moon surface conditions.

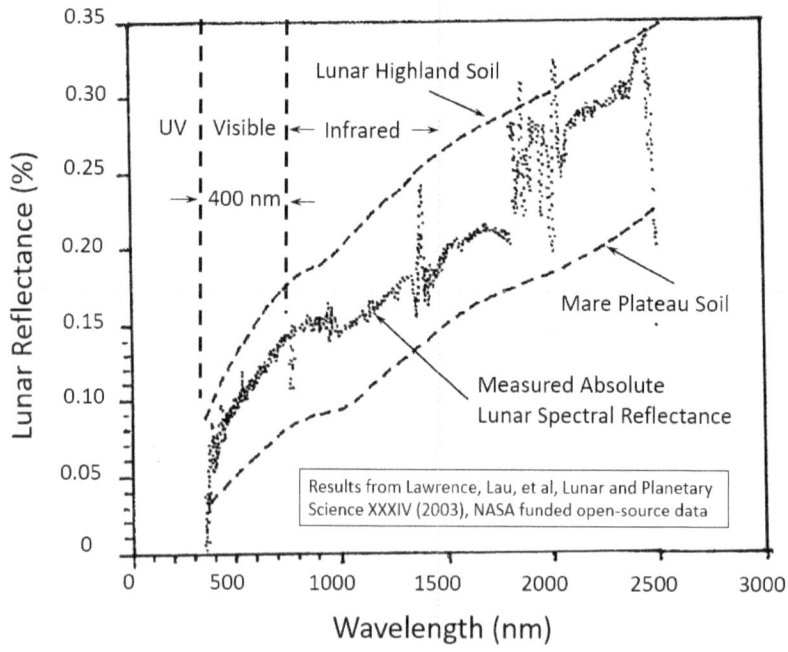

Chart of Lunar Reflectance vs. Wavelength

Results were published in "A New Measurement of the Absolute Spectral Reflectance of the Moon," S.J. Lawrence, E. Lau, et al, Hawaii Institute of Geophysics and Planetology, U. of Hawaii, Manoa, 2003.

In this paper Lawrence, Lau, et al compared the results of lunar reflectance versus wavelength for a variety of locations and also to the reflectance of lunar surface samples returned by Apollo astronauts. Measured reflectance from the moon and lunar samples showed a similar trend as a function of wavelength. In addition, results for the lunar samples provided upper and lower bound relative to reflected spectral data.

The human eye senses light wavelengths from 780 to 380 nanometers (nm = 1×10^{-9} m). Shorter wavelengths represent ultraviolet (UV) spectral range (UVA [400-315 nm], UVB [315-280 nm], UVC [280-100 nm] and longer wavelengths infrared (IR [780 nm to 2600 nm]). Radiation energy distribution in the solar spectral range skews toward ultraviolet and is visible as a function of the sun's equivalent radiant temperature (~5500°K).

The atmosphere reflects, scatters, and absorbs significant amounts of dangerous ultraviolet subject to environmental conditions. Solar radiation incident at the moon's surface does not experience atmospheric scattering, and hence is effectively identical to radiation in space. Overall distribution of energy from sunlight includes 5% UV, 42% Visible, and 53% IR.

Solar spectral shifting due to the heavens rending open and the resulting stretch of spacetime fabric would change light wavelengths in process from ultraviolet toward infrared. The magnitude of wavelength change would depend on the extent of the stretching.

A clue can be determined from the event recorded in the Old and New Testament prophesies. The removal of gravitational influence for distant mass resulting from rending the heavens would diminish vacuum energy density from far mass influence.

Suffice to say, the description in the above noted scriptural references of normally ultraviolet sunlight shifted to visible infrared indicates it will be consistent with a wavelength increase of about 300 nm.

Data published in the above technical paper shows lunar surface reflectivity change would be spectacular, ranging from two to four times greater reflected energy based on beginning and ending wavelengths. Though the lunar light reflected would shift to red, the amount of total energy flux reflected would increase several times based on the wavelength shift, giving rise to a blood red moon.

During the rending of the heavens, radiation including microwave, cell phone, and satellite communication will be distorted with a shift in frequency. Even the key fob remote and tracking devices will fail, at least for a time. As concerns the altered spectrum of light, by conservation of energy, heat transmission to the earth remains essentially the same.

When the heavens begin to open at the Lord's return, Jesus said events would unfold across the globe as lightning (Matthew 24:27, Luke 17:24). Light travels 300 million meters per second, but lightning travels about 120 thousand meters per second. Earth's circumference at the equator is about 40 million meters, which means light would circuit the globe in less than one tenth of a second, but at the speed of lightning it would take almost six minutes.

The distinction becomes important considering the result. At the speed of light, no one could register what had happened. Lightning however occurs quickly, but the outcome cannot be missed. With communication disrupted by the frequency shift, there would no prior notification around the world. But consequences of the event results in believers, both dead and alive, being removed to prepare for the time of God's wrath; "Yowm YHWH" or Day of Yahweh or Day of the Lord.

Lawrence Lau, et al recorded relatively low lunar reflectivity for rock and dust as a function of wavelength. In this study, reflectivity measured less than 0.05 (5%) in UV to almost 0.35 (35%) in IR.

The net effect of a cosmic shift from UV to IR (~300 nm) would be dramatic with the sky turning black across the earth due to a sun darkened like reddish coals in a campfire. Consequently, the moon would become deep blood red. As the fabric of the heavens rolls away, light from heavenly bodies (including stars) move across the sky, as Jesus said,

> For as the lightning, that lighteneth out of the one part under heaven, shineth unto the other part under heaven; so shall also the Son of man be in his day. (Luke 17:24).

Falling Stars

As noted, each star exhibits spectral characteristics of frequency distribution and intensity. During an event of rending the heavens and the resulting stretch, stars predominantly in infrared would shift to invisible long IR, visible spectrum to infrared, and ultraviolet to visible. Consequently, black skies still contain visible stars as the heavens are rolled away, giving the appearance of stars "falling."

This sequence of events unfolds across the sky and around the world. But what of places where the atmosphere is obscured by cloud cover? On an overcast day, when the sun moves low on the horizon and light passes through the atmosphere at a shallow angle, clouds frequently begin to dissipate.

This depends on atmospheric conditions, but as light at a shallow angle passes through more oxygen, UV is filtered and the remaining IR light tends to clear the clouds. In the case where all of the sunlight is shifted toward the IR range, it creates a powerful heater to dissipate overcast. Under normal atmospheric conditions, UV tends to increase humidity and convection, thus increasing cloud cover, whereas IR reduces humidity and clears the atmosphere.

Every person living during these events experiences trembling earth, blackened sky, darkening celestial lights, rolling back of the heavens and the appearance of Jesus Christ to the sound of a great trumpet and the shout of the archangel. Just as the heavens open, they quickly close again and nothing remains the same as God executes His final work.

Come Quickly, Lord Jesus!

Events disclosing the catching away of believers are outlined by numerous Scriptures. The Lord Jesus Christ promised to return in the heavens (Matthew 24:27 and 24:44). Paul the Apostle detailed how the Lord would descend with a shout, the voice of the archangel, and the trump of God to gather deceased and living believers (1 Thessalonians 4:16-17). In Acts 1:10, two angels declared to the disciples that Jesus would return as He left in the clouds. As quoted by Peter (Acts 2:17-21) regarding the last days, the prophet Joel prophesied,

> And it shall come to pass in the last days, saith God, I will pour out of my Spirit in those days upon all flesh: and your sons and your daughters shall prophesy, your young men shall see visions, and your old men shall dream dreams: And on my servants and on my handmaidens I will pour out in those days of my Spirit; and they shall prophesy: And I will shew wonders in heaven above and signs in the earth beneath; blood, and fire, and vapour of smoke. The sun shall be turned into darkness, and the moon into blood, before the great and notable day of the Lord come: And it shall come to pass, that whosoever shall call on the name of the Lord shall be saved.

Many published books and articles about the second coming of Jesus Christ argue these cosmic events in Matthew 24:29 represent Christ's physical return after the catching away to rule the earth a thousand years.

A fundamental problem with this notion rests on the virtually identical account at the opening of the sixth seal from Revelation 6:12-14, immediately followed by the Day of the Lord's wrath poured out on the earth. Clearly, signs at the opening of the sixth seal cannot be fit into any scenario associated with the Lord's triumphal return with the saints as described in Revelation 19:

> And I saw heaven opened, and behold a white horse; and he that sat upon him was called Faithful and True, and in righteousness he doth judge and make war. His eyes were as a flame of fire, and on his head were many crowns; and he had a name written, that no man knew, but he himself. And he was clothed with a vesture dipped in blood: and his name is called The Word of God. And the armies which were in heaven followed him upon white horses, clothed in fine linen, white and clean. And out of his mouth goeth a sharp sword, that with it he should smite the nations: and he shall rule them with a rod of iron: and he treadeth the winepress of the fierceness and

wrath of Almighty God. And he hath on his vesture and on his thigh a name written, KING OF KINGS, and LORD OF LORDS. (Revelation 19:11-16)

Revelation 6 and Matthew 24 as well as other Old and New Testament verses describe a series of signs associated uniquely with the rending of the heavens to display the return of Christ and catch away His people prior to the outpouring of the Day of the Lord's Wrath.

Subsequent events unfolding during the seven trumpets and the seven vials of wrath represent severe judgment upon unrepentant nations and absent the believers. During this period one hundred forty-four thousand Jews remain sequestered in the wilderness of Moab and Edom until the physical return of Christ on the ultimate fulfillment of Yom Kippur.

This appearing at the sixth seal represents a truncation before the end of the final 3-1/2 years for the sake of believers, as Jesus said,

And except those days should be shortened, there should no flesh be saved: but for the elect's sake those days shall be shortened. (Matthew 24:22)

After the catching away, during the seven trumpets and the seven vials of wrath, a few clues appear to indicate the time frame. After the fifth trumpet (Revelation 9:1) there is a period where men are tormented by the sting of locusts or scorpions for five months (Revelation 9:5, Revelation 9:10). Also, upon the sixth trumpet, four angels are loosed for an hour, and a day and a month and a year (Revelation 9:15).

These two amount to at least eighteen months following the catching away and prior to the fulfillment of Yom Kippur as described in Revelation 14:1-5 where the one hundred forty-four thousand come with Christ upon Mount Sion (or Zion).

As described earlier, the completion of Rosh Hashanah and the catching away should be before this eighteen-month window during the second 3-1/2 years of the seven-year *shabua*. Therefore, the Rosh Hashanah two years prior to the final Yom Kippur should coincide with the catching away of the believers. As noted earlier, that would be the "Day and Hour No Man Knows" because it would be declared by the

High Priest based on observation of the new moon by two witnesses. Pouring out of the seven vials or bowls of wrath then take place after fulfillment of the final Yom Kippur (Revelation 16:1-17) as part of the 75 (30+45) days preceding the celebration of Hannukah.

From Daniel 12:11-13, the Lord encouraged him concerning a hopeful event 1290 days after the Abomination and another at 1335 days. Since the final 3-1/2 years constitutes 1260 days, the remaining 30-day and 45-day periods are consistent with the 75-day period after Yom Kippur and preceding Hannukah at the time of Antiochus.

During this time the temple was cleansed, restored, and rededicated. To this day, Hannukah always falls 75 days after Yom Kippur. So also, by completion of the Day of the Lord, all of his enemies are to be vanquished, the Kingdom of David and Solomon restored in their completion, and the final Festival of Tabernacles satisfied in rest a thousand years, and final eternal rest in glory.

Festival of Tabernacles

To encapsulate the culture of harvest, God instituted a series of seven feasts outlined in Leviticus 23, grouped into three seasons or festivals. These extended harvest periods, from spring to fall, included key fruits and grains such as barley, wheat, oats, grapes, figs, pomegranates, and olives.

God called all males to gather with an offering at a chosen place (later identified as Jerusalem) three times in the year (Deuteronomy 16:16) marking harvest sabbaths of rest. Passover (Nisan 14), Unleavened Bread (Nisan 15-21), and First Fruits (Nisan 16) encompassed Passover Season, opening the harvest.

Fifty days (seven sabbaths plus a day) after Passover, the feast of Pentecost (Sivan 6) or shavout commenced, marked by a wave offering of two loaves of bread, symbolizing the entry of the Jews and Gentiles into faith. Pentecost memorialized receipt of the table of law by Moses, looking forward to the law written in the heart by the Holy Spirit.

A season of Tabernacles encompassed the longest period of assembly from Trumpets (Tishri 1) or Rosh Hashanah (meaning shout or alarm) to the Day of Atonement (Tishri 10) or Yom Kippur (meaning day of expiation) and finally Tabernacles (Tishri 15-21) itself or Booths (meaning tabernacle).

From the first day of the seventh month (Tishri, first civil month), to the final sabbath (Leviticus 23:39) on the twenty-second day after Tabernacles represented closure of the latter harvest. This eighth day, spoken of by the Lord, alludes to further rest after the millennium.

The feast of Tabernacles constitutes the only symbolic festival carried out during the millennial reign of Christ (Zechariah 14:16), which represents a looking forward to final rest in God's presence in the heavenly realm.

In addition to the seven Feasts of the Lord, Scripture and Jewish tradition include others such as Tisha B'Av (Av 9), Hanukkah (Kislev 25 – Tevet 2/3), and Purim (Adar 14). Jewish festivals instituted by the Lord in Leviticus 23 reflect the call of Jesus in John 4:

> Say not ye, There are yet four months, and then cometh harvest? Behold, I say unto you, Lift up your eyes, and look on the fields; for they are white already to harvest. And he that reapeth receiveth wages, and gathereth fruit unto life eternal: that both he that soweth and he that reapeth may rejoice together. And herein is that saying true, One soweth, and another reapeth. I sent you to reap that whereon ye bestowed no labour: other men laboured, and ye are entered into their labours. (John 4:35-38)

Return to Normal?

After the catching away of believers, what next? Does the cosmos and the world return to some semblance of normalcy? Just as the heavens rolled back to herald the appearance of Jesus Christ, the cosmic fabric again closes. At this, light from the sun, moon, and stars return to their former state, but nothing else resembles normal. After unprecedented events during the sixth seal, all people remaining cry out:

…Fall on us, and hide us from the face of him that sitteth on the throne, and from the wrath of the Lamb: for the great day of his wrath is come; and who shall be able to stand? (Revelation 6:16,17)

Suddenly multitudes clamoring against God and the faithful in Christ will get what they demanded. With true and faithful believers gone, the world senses foreboding and a harbinger of destruction to come. Beginning with the seventh seal, the ensuing seven trumpets, and the seven vials pour out the wrath of God upon the disobedient and rebellious.

Attempts at Israel-Palestine peace agreements the past fifty years include Camp David Accords (1978), Madrid Peace Conference (1991), Oslo Accords (1993, 1995), Camp David Summit (2000), and Abraham Accords (2020). Negotiations in 2025 seek to expand particants to the Abraham Accords in hopes of a settlement to Israel-Arab conflict. The current Abraham Accords, patterned after Abraham's covenant with Abimelech at Beersheba (well of the oath, well of seven), lay the framework for the final covenant of seven spoken of by Gabriel (Daniel 9:27).

At some point in the near future, this framework covenant will serve as basis to constrain Israel under an agreement to be confirmed by Antichrist. As prophesied by Isaiah, this will be a "covenant with death" or "agreement with hell" (Isaiah 28:15-18). The Lord reveals his work in an appointed time, though clearly the time is short.

After revealing of the Antichrist figure, and confirming a covenant with Israel, those who misinterpreted the message will require mental gymnastics trying to process how they missed the obvious. Jesus will appear suddenly at a time appointed by God Himself. As in the parable of the virgins, "…at midnight there was a cry made, Behold, the bridegroom cometh; go ye out to meet him." (Matthew 25:6). Time to get ready. Even so come, Lord Jesus.

King Solomon, author of Ecclesiastes closed by observing:

The words of the wise are as goads, and as nails fastened by the masters of assemblies, which are given from one shepherd. … Let us hear the conclusion of the whole matter: Fear God, and

keep his commandments: for this is the whole duty of man."
(Ecclesiastes 12:11,13)

Young Tom Cox with Dillard Lutes on Horseback

Chapter Eighteen - Final Curtain

TOM COX RECALLED late rides home on horseback as a boy, clinging to the back of his father, Dillard Lutes. The obituary for Dillard Lutes (1860-1898) states he was shot and killed in a corn field after a card game in Pine Grove, Kentucky, just west of Winchester.

Though it mirrors the game of Clue, this outcome became deadly serious. As a result of his father's untimely death, Tom Cox, my wife Debby's grandfather refused to allow card games in the home. In the process of doing family genealogy, Debby received a copy of a photograph of Dillard Lutes. A century before, someone used his picture for target practice.

Games favor varying levels of both skill and chance, but do not allow purpose; neither prescribed outcomes of winning nor losing, unless you are in a casino playing against the house. In a game of pure chance, throwing a fair die is regarded as a simple probabilistic event where the outcome is one of six possible numbers. That means nothing skews the result to favor a given number.

For a large quantity of throws of the die, each number can be expected to come up equally. If someone or something favors a different intent, then purpose precludes pure chance. Chance can then be characterized as a purely random event, or nothing: no purpose, no control, no outside influence.

In a Sunday School class someone once questioned, "Why don't we see miracles happen like in the Old Testament?"

My immediate response, "The reason is because you ask the question."

A skeptic once joked dismissively, "The lottery winner always believes in miracles." More correctly, "A believer does not look for miracles, they follow." Life can appear unfair, routine, or selectively favored. Ultimately, purpose percolates up from both good and evil. Every person occupies a small piece of space and time, hopefully to the realization that there exists a larger plan and purpose. All face certain but diverging consequences, but in the end God has His finger on the scale.

What You Demanded

In the prairies of South Dakota exists a desolate wasteland called the Badlands Bombing Range. During World War II a section of the Pine Ridge Indian Reservation received a shipment of derelict automobiles to be used as targets for B-17 bomber training. Now little remains except for skeletal remnants of those target vehicles and unexploded ordinance overgrown with buffalo grass. During its heyday, the warning was, "You Don't Want to Go There." But, if that is what you demand, be forewarned, no one leaves with T-shirts or bragging rights. Some places are not destined for living.

Derelict Vehicle Targets in Badlands Bombing Range

Nearing the end of the forty-year trek from Egypt and approaching Jordan, Moses spoke at length in Deuteronomy to the people of Israel. He rehearsed events of the exodus, requirements of the law, and warned of blessings for obedience and curses for disobedience.

One caution related to the choice and nature of a king (Deuteronomy 17:14-20). Over two hundred years later, the people, under Samuel as judge, demanded a king like every nation. This deeply troubled Samuel, yet God authorized him to give the people what they demanded while admonishing them the consequence of autocratic rule rather than the judicial theocracy (1 Samuel 8:1-22) under God's rule.

In response to the demands of the people, God led Saul the son of Kish to Samuel who anointed him king and revealed Saul to the people. For a time, the kingdom functioned well. Saul and the people even defeated the Ammonites which helped cement his rule. Then the kingdom began a slide toward cautioned authoritarianism under Saul, disobedience regarding destruction of the Amalakites, and a breech with Samuel. In turn, God directed Samuel to anoint young David (son of Jesse) as king instead.

Principles in Scripture can be understood by the meaning of

associated Hebrew words. The name of Saul is no exception. When the people demanded a king, God gave them a man named Saul (meaning to ask for). The name Saul and Sheol both have origin in the Hebrew word *shaal* or demanded.

The name David or beloved reflects God's purpose of a seed of the tribe of Judah carrying the promised blessing until the rightful heir, Jesus Christ, came into His reign. The concept of Sheol (underworld, hell, grave) reflects the consequence of choices. Some ask critically how a loving God can send people to hell.

That question, usually framed as a challenging accusation, seems an irresolvable conflict to the nature of God. The biblical concept of Hell echoes the warning that parents give concerning the local bar (or the Badlands Bombing Range), "You don't want to go there."

Sometimes curiosity and temptation take root; just to check it out. The outcome can be a lifetime of regret. Hell is a place you don't want to go, not a threat. It is a place of separation from God; don't go there.

Disobedience and evil cannot exist in the presence of God; therefore, He created a place of separation, or *Sheol*, implying that's "what you demanded". Insisting on your own way or something other than the plan and purpose of God can mean He gives you what you want, saying, "That's what you demanded." It was not His plan. No wonder the Apostle Paul adopted his Hellenized name rather than his Hebrew name, Saul; Paul sounds much better than the implied *Sheol*.

There is no good thing exempt from Satan's ability to distort or manipulate for evil, when demanding our own way. So also, there is no evil outside the reach of God's mercy, when choosing His way. He can and will use what you demanded, to His purpose, subject to faith and repentance.

All sin and rebellion directly follow the rejection of a personal relationship with God. Sin reflects a symptom of the larger problem, the loss of life's foundational relationship.

Thank God there is coming a rest when God's purpose on earth will be complete. That time will be a personal rest for believers as well as rest for the earth and a government where believers rule and reign with Christ for a thousand years (Revelation 20:4-7), and finally spend eternity in glory.

The principle of what you demanded has implications to career choices, relationships, even the outcome of elections. Increasingly voters select leaders, like Saul, who turn their back on God. Samuel rehearsed the consequence to the people of Israel, who replied, "Nay; but we will have a king over us…" (1 Samuel 8:19).

The name Saul vividly illustrates the outcome. Be careful what you demand. A further irony of Saul's life and disobedience unfolds with his failure to execute judgment fully on the nation of the Amalekites (1 Samuel 15).

Decades later, at Saul's final battle with the Philistines on Mount Gilboa, an Amalekite descendant arrived to plunder the slain army. Finding Saul dead, he plotted to lay hold on his crown and bracelet to prove to David that he had killed Saul, a fabricated ruse. Of course, David used the claim of the Amalekite to pronounce sentence against him and have him slain rather than levy a reward.

For believers, the consequences of faithful obedience and service, like talents for the master's stewards will shortly become clear. Jesus Christ comes quickly to rend open the heavens beginning a final chapter of His kingdom.

The Land of NOD

Swirling specks of matter,
descend through rays of light.
Ghosts of ancient treasures,
possessing all in sight.
Till dust encase and dross encrust,
the soiled and littered scene,
invoking some redeemer,
to deliver the unclean.

In a cunning psyche,
the battle plans foment,
as skillful hands prepare themselves,
the scheme to implement.
With careful crafted tools of war,
refined in sore travail,
now arrayed against the foe,
till Neatness does prevail.

How oft we gather objects,
to use or just admire,
but ne'r define a resting place,
to store what we acquire.
With time the goods accumulate,
piled high in heaps and drawers,
to frustrate those from finding,
the things they're looking for.

How much more of value,
investments planned for space,
and built into a construct,
where each thing has its place.
So chaos defers to structure,
as darkness succumbs to light,
till Order lays a framework,
which is pleasing to the sight.

Every earthly object,
cries with cryptic voice,
if shipwrecked on a foreign shore,
and not its' port of choice.
Why mortgage future moments,
to rescue what's misplaced,
by careless navigation,
within this sea of space?

Rather forge with diligence,
a plan to sculpt the will,
in seeing each exertion,
its' ending to fulfill.
Till training and rehearsal teach,
the worth of each detail,
accomplished without falter,
in Disciplined travail.

So do not take the easy course,
where signs do not portend.
The way that seems much smoother,
'tis harder in the end.
But in the Land of NOD,
there's none who disagree,
Neatness, Order, Discipline,
the import of these three.

William McGreehan (1991)

King and Jester

The pediatrician at the University of Florida Shands Hospital in Gainesville, Florida pronounced, "That's amazing! She now appears to have hip sockets." In 1974, while ministering at the WMO center in Gainesville, our newborn daughter, Dawn, received complimentary care for many months due to her lack of hip sockets, a congenital condition.

Months earlier the pediatrician warned she might need surgery, therapy, and possibly never be able to walk. After the removal of her awkward braces, the pediatrician could no longer move the hip joints out of place.

We mentioned to the doctor that our ministry team had prayed for a miracle that morning after months of unsuccessful treatments. Asked if he believed in miracles, the pediatrician quipped, "Occasionally, we see what is called, 'Spontaneous Healing.'" A label helps to conveniently pigeonhole the obvious.

Over fifty years ago, after coming to faith in Christ, the ministry presentation at Penn State raised questions about the observations of science and the miracles recorded in Scripture. At the time, issues such as the origin of the universe and the genesis of life seemed secondary to a relationship with God.

Now, with the culture, nation, and world on the verge of moral implosion, answers to those questions are foundational to the framework of life and practice. As an engineer and a builder, experience demonstrates the integrity of a structure depends on solid footing.

That explains why Satan attacks the foundation of the culture and the church. Yet God's kingdom and His Word stand firm on the work of Jesus Christ, the Chief Cornerstone. In the final analysis, the work of Satan serves only as a fitness center for resistance training to the believer.

When in 1971 Don McLean's "American Pie" looked back over the culture, he composed the song as a series of musical laments, electing to dismiss hope. By contrast, Abraham prepared to offer his promised son, looking ahead foreseeing God would make provision to raise him up.

By faith Abraham, when he was tried, offered up Isaac: and he that had received the promises offered up his only begotten son, Of whom it was said, That in Isaac shall thy seed be called: Accounting that God was able to raise him up, even from the dead; from whence also he received him in a figure. (Hebrews 11:17-19)

Court Jester, the Welcoming Committee Chairman

Throughout history as well as biblical narrative, a curious relationship exists between the splendor of royalty, the court, and subjects. Instrumental in the theatre of kingdoms, a unique role arose to entertain and manage imperial guests. To lighten the pompous atmosphere of court and control pretentious ministers, a jester or joker received employment. Often small of stature, outrageous personality, and outlandish attire, the court clown's primary responsibility fell to amuse and preoccupy players; the proverbial wild card.

A favorite jester tactic involved poking satire at formal proceedings, ministers, and wives. Traditionally, the king and royal family remained off limits, though human nature being what it is, the joker occasionally overstepped bounds and wound up unemployed—or worse, on the chopping block.

From royal theater games developed such as chess or a deck of cards, each built around a cast of characters, a playing field, a set of rules, strategies, and outcomes to win or lose, notwithstanding chance and skill. One constant always remained; tension between imperial authority and a wild card, such as "pawn promotion" in chess.

American Revolution Chess Board and Players

The Kingdom of God portrays a scene set on a grand stage, conceived by a creative Author, assembled by a Master Builder, populated by a vast array of costumed players, each with particular vulnerabilities.

As the story plays out and the plot develops, surrounded by intrigue enters the pretentious jester Heylel, bound by a singular purpose: to promote resistance, and in process of time, distract the attention of all present from the underlying plan.

Characters weave in and out of the story as events unfold, all taunted by the jester; a few departing the scene celebrated. Most depart unceremoniously, reciting lines but never understanding the purpose. Sadly, many participants neglected to read and interpret the script; others informed, misunderstood the meaning.

Suddenly, on the cusp of chaotic rebellion, as the jester attempts to lay hold on the crown, curtains come down and celestial lights dim. Distracted, players exit the stage as the realization settles in; it is not a board game, international intrigue, nor theater, but a reset of history as we know it. Ultimate Deus ex machina on a cosmic scale; an improbable resolution to an impossible situation as the Author Himself briefly reenters the stage to foil the joker and complete the supreme purpose: redemption and judgment according to His eternal plan.

...Surely I come quickly. Amen. Even so, come, Lord Jesus. The grace of our Lord Jesus Christ be with you all. Amen. (Revelation 22:20, 21)

Fissures in Stone

Under a sultry August sun,
Faced with tasks I seek to shun,
Duties call and demands increase,
My strength wanes
and I long for release.
The moment comes
when I make my escape,
Off to the wood, a short walk it takes.
Wandering on I find a cool grove,
Where under the
bows a rocky crag shows.

Into the shade a curious form,
Gradually appears from
imposing gray stone.
Hidden from view of impatient eyes,
A small dark
fissure my attention spies.
There at the base I am inexorably drawn,
To a small open space,
the way barely shown.
With abandon of
youth and caution aside,
I slowly crawl to a world denied.

There as light and warmth abate,
Replaced by
strange and unseen shapes,
A curious pleasure envelops my mind,
Experiencing sensations
reserved for the blind.
Oft in my youth I followed this lore,
Always drawn deeper
in desire to explore.
Till caution taught
young bested my need,
To find out where
this darkness did lead.

Now grown older I'd forgotten the cave,
Till reading of those
who descended to save,
Some poor soul caught fast below,
Searching for pleasure, no light in tow.
Recalling the fond excursions of youth,
I took the lantern that shone in truth,
To find that place where many a day,
I 'scaped from toil and slipped away.

There now open and exposed to view,
The path below trod not a few.
What was once a secret haunt,
Has now become a quest to flaunt.
Searching below with piercing light,
Mortal dangers that hid from sight,
Are suddenly
become an unwelcome find,
And made to shame the foolish mind.

Casting the light to darkness ahead,
Fills my heart with fear and dread.
For there a deep, unwelcome tomb,
Opens wide to certain doom.
With cautious gaze I plumb the depth,
For unwary souls
gone to gruesome death,
And there I see the countless lost,
Who my paths have daily crossed.

Stepping back from a scene so sore,
My heart is pained for many more.
And I resolve to find a way,
To duly warn the careless prey.
Retreating from the "what if" past,
I carved in stone a note to last.
And reach the mind that wanders down,
To 'scape the trouble all around.

"Many strong men and wise men too,
have entered here as now you do.
Shunning light they groped as blind,
Until a sudden death they find.
So attend to
Scripture where God before,
Gave wise and
prudent direction of yore.
Walking only on lighted ways,
They grew in Christ with passing days.
But unbridled minds
like fissures in stone,
Lead to pain and loss unknown.
Though enticing and harmless
the entrance is lit,
Beneath the surface
descends to the pit."

William McGreehan (1991)

Rend the Heavens,

...and Come Down.

www.ingramcontent.com/pod-product-compliance
Lightning Source LLC
Chambersburg PA
CBHW060416130626
46555CB00005B/2087